BEYC EINSTEIN'S UNIFIED FIELD

Gravity & Electro-Magnetism Redefined

John Brandenburg Ph.D.

Other Books of Interest:

LIFE AND DEATH ON MARS
THE CASE FOR THE FACE
EXTRATERRESTRIAL ARCHEOLOGY
INVISIBLE RESIDENTS
THE COSMIC WAR

BEYOND EINSTEIN'S UNIFIED FIELD

John Brandenburg Ph.D.

Adventures Unlimited Press

BEYOND EINSTIEN'S UNIFIED FIELD

by John Brandenburg Ph.D.

Copyright 2011

ISBN 13: 978-1-935487-42-5

All Rights Reserved

Published by:
Adventures Unlimited Press
One Adventure Place
Kempton, Illinois 60946 USA
auphq@frontiernet.net

www.adventuresunlimitedpress.com

Picture credits: All images taken from Wikipedia Commons unless otherwise marked.

TABLE OF CONTENTS

Prologue: Einstein's Final Triumph..........................3
Chapter 1. Mars Hill: the Cosmos as It Is..........21
Chapter 2. A Book of Numbers and Forms......43
Chapter 3. Kepler, Newton and the Sun King....65
Chapter 4. Magnus and Electra............................81
Chapter 5. Atoms of Light....................................103
Chapter 6. Gravitas: Einstein's Glory..................121
Chapter 7. The Aurora...147
Chapter 8. Tesla's Vortex — the Cliffs of Zeno....175
Chapter 9. The Hidden 5th Dimension.................195
Chapter 10. The GEM Unification Theory............207
Chapter 11. Antigravity and Human Flight..........235
Chapter 12. Alpha and Omega: the Cosmos.........257
Epilogue: The Summit of Mount Einstein.........273
Afterword: Summary of Results..........................287
Chapter Notes..289
Chapter Bibliography...301
Appendix..305
Index..333

Dedication: To Dr. John T. And Muriel A. Brandenburg, my beloved parents, who always encouraged me.

Prologue: Squaring the Circle, Einstein's Final Triumph

"The way of Heaven is round, and the way of Earth is square."
Old Chinese proverb

"The word that can be spoken, is not the perfect word;, the way that can be walked is not the perfect way."
Lao Tzu

"There are only atoms and the void."
Democritus

Beyond Einstein: GEM Unification

Einstein triumphant.

Einstein was right in his final hypothesis, that electromagnetism and gravity could be unified with a result useful to humanity. In this he was joined by Tesla, who also believed EM and gravity could be unified, and had been a hidden influence in Einstein's life.

Einstein may have been wrong about some of the details of such a theory, but his basic premise and intuition were correct: that the two long-range forces of nature were a set complete largely within themselves, like children of the same parent. Despite the fact that Einstein failed to publish a successful unification theory, unification has been accomplished using the pieces that he created. Thus, it is his final masterpiece, accomplished long after his death, but still his ultimate vindication. For he was a man who had been right about so many things, even if he sometimes thought he was wrong. Wherever he is, he must derive some final satisfaction from this event. One can imagine that he sits now at the summit of Olympus, where Heaven and Earth come together in the court of the gods. It was a long hard

Beyond Einstein: GEM Unification

climb to this summit, after many failed attempts. Indeed, many disdained him in the end for even trying, but this final ascent is all the sweeter for its views from the summit.

In this book we will explore this new theory of unification of two of the most powerful and far reaching forces of nature, and its meaning: that it also unifies Heaven and Earth. We will discuss how much of this unification Albert Einstein anticipated and laid the groundwork for, and how much came from other workers of vastly different intuition and viewpoints, especially Nikola Tesla. It will be discovered that this unification involves the mysterious number 42.8503..., that it discovers a third field that is intermediate between gravity and electromagnetism, and that the unification radically changes our view of the cosmos and how it operates.

It is the intention of this book not to describe the theory as final product, because it is only a newborn child, but rather to describe the journey through the ages that led to it, and what appear to be the theory's major implications. It will be a tale of how each part of the puzzle of the two forces was separated and carefully studied apart from the others, and finally laid into place in the grand picture. It is by this process and telling the tale of the discovery that the meaning of the final unification will be made more full to the reader, and its future development envisioned. The goal of this work is to convey understanding, not just tell one how to build a starship.

Beyond Einstein: GEM Unification

Nikola Tesla, who laid the groundwork for unification and molded Einstein's life.

To understand a thing fully one must understand its poetic as well as scientific truth. For this reason the book will concentrate on imagery and geometrical diagrams rather than numerical and algebraic manipulation. It will also talk about people and their ideas and the context in which they thought and spoke, more than the formalism they used to express these ideas. However, part of the joy of science is mathematics and numbers, and so at the end of the book mathematical chapter notes are included for those who share my intoxication with these methods, and who are able with their calculators. An appendix is found at the end of this book describes the unification theory in mathematical detail for those who have not read the journal articles describing the GEM theory. Finally, after exploring this island of new understanding, we will, in the words of Isaac Newton, walk the new shoreline of wonder. For the solution of each

Beyond Einstein: GEM Unification

mystery has, in its core, the discovery of a deeper mystery, and that is the joy of science, that each doorway passed leads to a corridor of new doorways. The scientist's vocation is to find the keys to these doors and to affect the passing through of them by the human caravan. For with each door passed, we realize how we are changed for the good by this passage, all of us, so that we would never retreat from the new place we behold for all the treasures of the galaxy.

Ancient Chinese coin depicting the square Earth surrounded by a circular Heaven.

Four great forces shape the reality that we live in: gravity, electromagnetism, the strong nuclear force, and the weak nuclear force. Each force has its own realm where it reigns supreme and each plays its role in shaping our life. Gravity and electromagnetism are forces with infinite range. Gravity shapes and moves the cosmos to the greatest degree. It is the product of a large numbers of particles, hence its connection with entropy or disorder. Next to gravity is the other powerful long-range force of the cosmos, electromagnetism, which we have made our servant. By it

Beyond Einstein: GEM Unification

the atoms that make our being are held together. For, by electromagnetism, we perceive and measure the cosmic realm where gravity reigns and see it display its awesome strength. And here is a riddle: *Would the stars move by gravity if we could not behold their light?*

Medieval Cosmology: Heaven and Earth meet.

Next in the list of forces are the short-range forces of nature, the strong nuclear force and the weak. They are limited in range but their action is mighty. The strong force, as it is known commonly, is so named because it holds together the positively charged particles of the nucleus, which would fly apart due to electromagnetism if the strong force did not exist and exceeded it in strength. Hence its name, for it is stronger at short ranges than electromagnetism. It has been described as a "nuclear glue" because like glue it only holds particles together that "touch" each other, that are separated by only a nuclear particle radius, whereas gravity and EM forces pull and push particles at distances much larger than the nucleus. Compared to the strong force the gravity and EM forces

Beyond Einstein: GEM Unification

appear as true force fields, full of dynamics and rich behavior, whereas the strong force seems primitive and underdeveloped. However, this perception, we shall see later, is an illusion brought about by the scale in which we live. The strong force, it turns out, has a rich and splendid reign in its own castles of the nuclear particles themselves, and what we see of the strong force in our plane of existence is only the faint glimmers of the strong force that escape from the castle's shuttered windows. Suffice it to say for now, that because of the existence of strong force we exist, made of atoms more complex than hydrogen, and we are warmed by the Sun in the daylight and enthralled by the starlight at night.

Beyond Einstein: GEM Unification

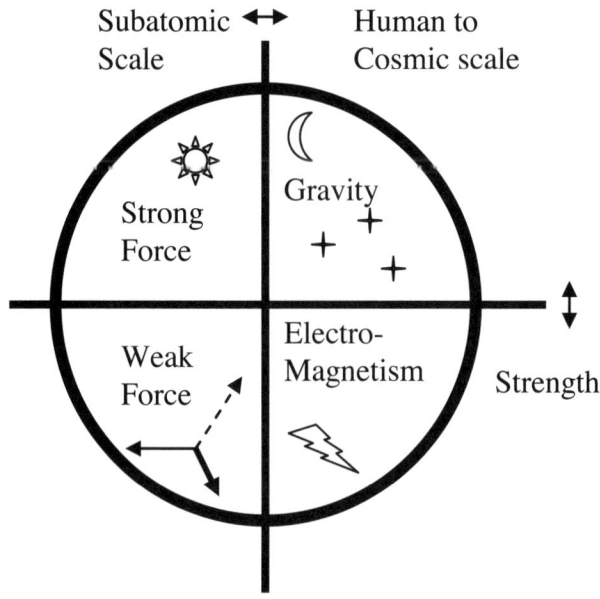

The four known forces of nature, divided by scale and strength of action.

The weak force is even more mysterious than the strong in terms of our everyday experience. One would ask, if the weak force disappeared would anyone notice. The answer, suffice it for now, is that we would definitely notice its absence, as half the nuclear particles in our bodies would instantly explode. It is the subtlest of the forces at our plane of existence, but powerful. It can be viewed as the helper to the strong force in the realm of nuclear particles. The weak force, in its simplest description, governs radioactive decay. Despite its mysterious nature, a narrow mathematical bridgehead has been established between the weak force and electromagnetism which is enlightening and enthralling. We will explore this in later works. Suffice it for now that strong and weak forces are short range, and only become active at

Beyond Einstein: GEM Unification

the distances of the radius of a nuclear particle and smaller, and can be considered a couple. We will consider them in later works. In this book, the two long-range forces of nature, gravity and electromagnetism, will hold our attention. Together these long-range forces rule the dynamics of our plane of existence, and beyond that, the grand cosmos above our heads.

Here is a mystery: we shall discover, later, that electromagnetism is not one force, but actually three, and this will be a crucial truth. The unification of gravity with electromagnetism will be accomplished because of a force field that is an intermediary field between them, part of both. This is the third force of electromagnetism that we will discuss. For now, think of electromagnetism as a unity of one. These are the two great forces, gravity and electromagnetism, coupled in their breadth of action, whose unification will be the goal of our journey.

Part of the mystery of the cosmos is that it can be understood a piece at a time, in a succession of useful approximations, just as the longest journey is a succession of steps.

All journeys begin with a single step. Einstein was a man who dared to take that first bold step. He was born in Germany in March 1879 but his family was forced to move a few weeks after his birth. His father's business was failing due to the work of Nikola Tesla. For his father had helped found an electrical business based on the direct current power distribution concept of Thomas Edison, but Tesla's alternating current system had triumphed. In these days Europe was moving from the era of gaslight to the electric light. The result of this family disaster was to dominate Einstein's childhood. So Einstein's early life existed in the long shadow of Tesla.

Beyond Einstein: GEM Unification

Einstein came to manhood at the foot of the Alps, in Switzerland and Northern Italy, in the German-Jewish community of central Europe in the late 1800s. All during his childhood he would gaze at the far off peaks wrapped in snow year round, so much higher and colder than the warm valleys he lived in. To gaze at those summits was to grasp how two planes of existence could exist side by side and yet be alien to each other.

Later in Princeton, in the United States, in the final stages of his life, he continued to work on his theory to unify electromagnetism and gravity but could produce nothing that matched experiments. He was now an exile from his homeland, and an outcast in his field of physics. He had gone from the leader of the Olympic Institute, an informal discussion group of dreamers that he organized while a lowly patent examiner, to being the director of the Kaiser Wilhelm Institute in Berlin, the world's foremost center of physics, and now to the Institute for Advanced Study. He had made the full circle of the heavens, seen every glory of them, but now he was once again a rebellious dreamer on the outside of the physics community. The quest for a unified field theory of gravity and electromagnetism had consumed him and his career.

Einstein became a legend in his own time, and his quest for unification was mythic. He had joined Tesla in the dream of gravity controlled by EM fields and like Tesla had embarked on a long, lonely, journey to oblivion. To the public Einstein was still the greatest physicist who had ever lived, but to the physics community he was now regarded as an old crank, and his proposed unification was not mythic, but a myth. It had become, for Einstein in his fading years, the ultimate quixotic journey. Schrodinger, his old friend, when asked about Einstein's quest for unification proclaimed him a "damn fool."

Beyond Einstein: GEM Unification

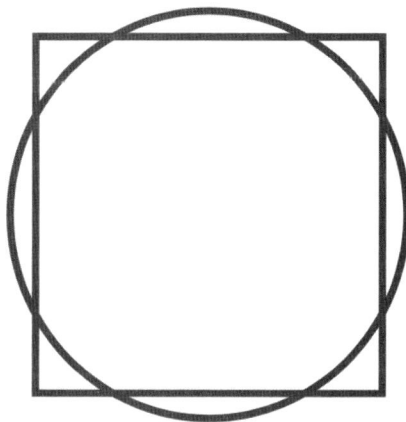

The squaring of the circle: the unifying of Heaven and Earth.

Einstein, by his stature, and by his failures, had transformed the quest for the unified field into the legendary Greek problem of "squaring the circle," an ancient mythic quest. It had been long a challenge to draw a square of exactly the same area as a given circle using a straightedge and compass. The problem could not be solved, because π could not be known exactly and thus could not be constructed geometrically. So it was dismissed as one of those old problems without solution. To try it was to fail; to persist in it was folly. It was plain to the physics community at the time that the path of advancement, excitement, and government funding was quantum physics in the subatomic scales, a path Einstein had rejected. To him the answer lay not in the subatomic, but in the cosmic scales. So he labored on, increasingly isolated, until the last days of his life, dismissed by his colleagues even as he was revered by his public. The fields of EM and gravity could not be unified, it

Beyond Einstein: GEM Unification

was said, just like the circle cannot be squared. But the circle can be squared, as it turns out.

To square the circle can be a metaphor for solving a problem that contains the world. The circle represents nature, the unknown, the void. The circle represents the cosmos as it is, free from our conceptions and remote from our everyday world. To square something is to understand and rationalize it, to "rectify it," to "make it right." To make something understood in the context of known facts is to "square it with reality." Therefore, to square the circle means to understand the sky and what it contains, the stars, galaxies, and the UFO.

To square the circle one must find the magic angle; we will call it delta, Δ. Find that one magic angle, Δ, and the height H of the right triangle it defines gives the side of the square that squares the circle. To do this one must give up a precious human thought, that of achieving perfect exactitude and ultimate knowledge. The circle cannot be squared exactly because π cannot be known exactly.

Beyond Einstein: GEM Unification

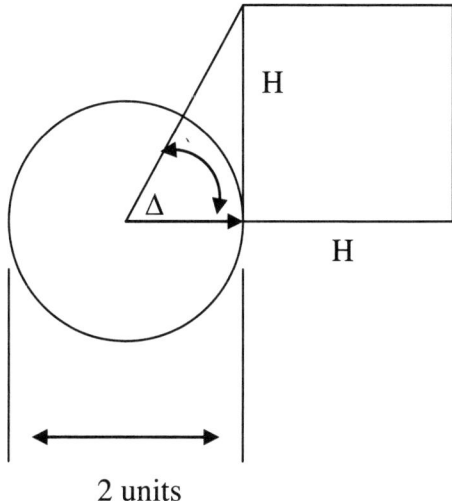

2 units

The angle that squares the circle: Δ, giving H.

However, the number π is a very useful number even if it is not known exactly. It can be known only by successive approximation. Exactitude is the goal of philosophers, but not of carpenters and others of those who build. So it is even with physics, that we must solve problems continually based on partial knowledge and so achieve good approximation. In truth, no perfect compass, no perfect straight-edge, no perfect sheet of paper, exist, we can only achieve accuracies of a part per hundred with such things as we have. So loosed from the burden of the unattainable we can square the circle to good approximation. On one level, it matters not how we find Δ, it only matters if it works. We find Δ here by dream analogy.

Beyond Einstein: GEM Unification

1. Make a wheel one unit across
2. Roll it one revolution and call this distance a "cycle." The cycle is now π units.

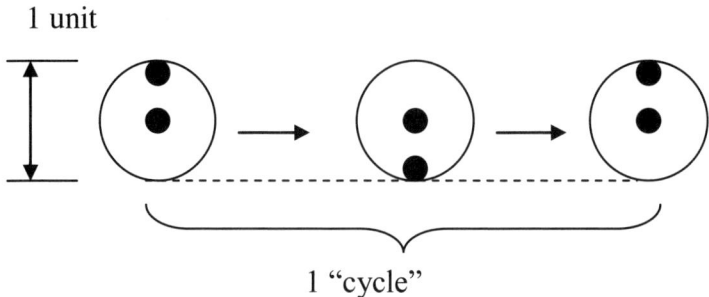

3. Now cast a ring of gold with a circular cross section 2 cycles in diameter and with a total outer diameter of 8 cycles with an inner diameter of 4 cycles. This forms a

torus.

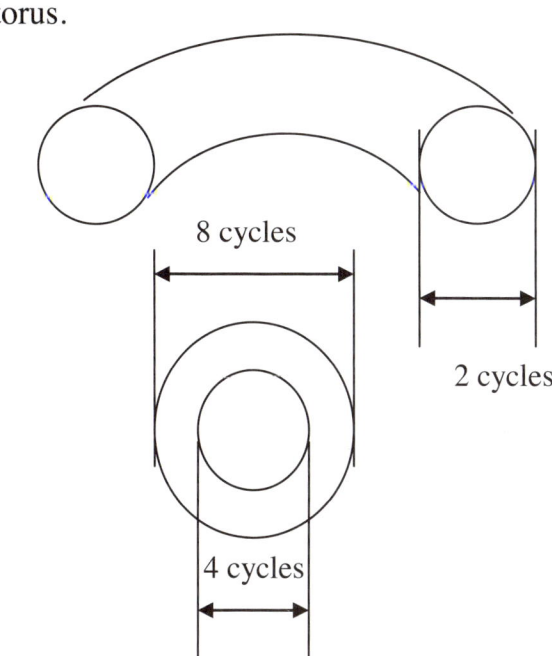

4. Now hammer the gold ring into a square plate 1 unit in thickness. Its sides will be 42.8503 units in length. We will call this number sigma, σ. Remember it well, for we shall see it again on our journey.

Beyond Einstein: GEM Unification

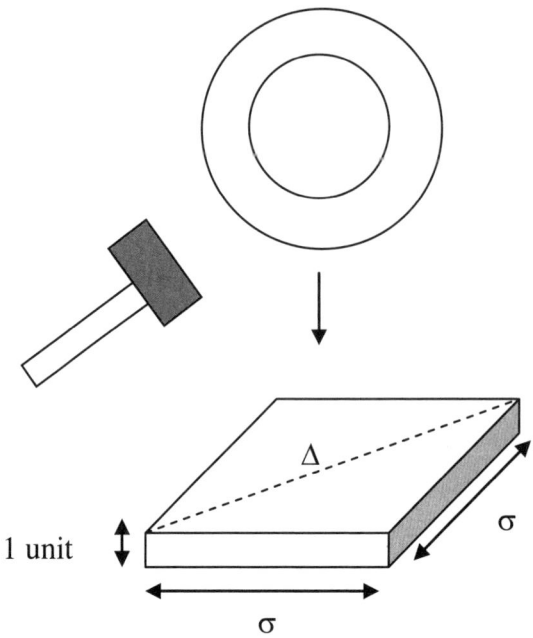

5. Take the diagonal of the square formed from the ring; this is Δ.

6. Draw a circle of unit radius, divide the circle into 360 degrees after the custom of the Chaldeans. Then draw a right triangle of base 1 unit and base angle on the circle of Δ degrees. The height H of the triangle is then the side of a square of area equal to that of the circle to 2.5 parts per thousand.

Thus is the circle squared by a dream.

So it is that the circle can be squared, not perfectly, for perfection lies not with humanity, but to good approximation.

Beyond Einstein: GEM Unification

Thus, the myth is but a myth. The circle can be squared, it can be squared many ways to better approximation. Once one gives up the burden of exact solution, everything becomes possible, even in some cases an exact solution by accident. We have accomplished everything we do by approximation. We have crossed oceans, mounted the sky, and journeyed to the Moon and back six times and sent four probes on interstellar flights. All of this was done by approximations. So, if the circle can be squared, the fields can be unified, also to good approximation. If one can square the circle, one can construct a building; if one can unify the fields, one can build a starship.

So our journey begins, to scale the faraway summit of the Mountain of Unification, to stand in the place where the square of Earth meets the circle of Heaven. That place that Einstein dreamed of in his Olympic meeting room.

So my fellow journeyers, put on your sturdy Swiss hiking boots, and your warm Swiss hats. We prepare to set out upon our great journey, to the icy summit in the distance. This is the summit of Mount Einstein, named for the goal he sought so long, which rises from the fruitful checkered plain of physics, which was, before his arrival, a wilderness.

Beyond Einstein: GEM Unification

Mount Einstein, at whose summit gravity and electromagnetism are unified.

Many paths lead to a mountain top, but some are more treacherous and difficult than others. We will choose a good path to the summit as we near the mountain's base.

Let us be off now, down the road, as we are promised blue skies if we do not tarry. A great journey begins with a single step, so let us boldly step forward down the road.

Chapter 1. Mars Hill : the Cosmos As It Is

"To acquire knowledge, one must study, but to acquire wisdom, one must observe."
Marilyn Von Savant

"Know the smallest things and the biggest things, the shallowest things and the deepest things. As if it were a straight road mapped out on the ground, ... these things cannot be explained in detail. From one thing, know ten thousand things. When you attain the Way of strategy there will not be one thing you cannot see. You must study hard."
Miyamoto Musashi, *The Book of the Five Rings*

From the diary of the expedition

Sometimes it is best to ask, 'how', rather than 'why.'

As we prepare and equip ourselves for our journey, we tarry in Athens, and we amuse ourselves by going to Mars Hill or the Areophagus, where the great philosophers are known to congregate. The Greeks were not known for their

Beyond Einstein: GEM Unification

experiments; however, they were keen observers. We encounter there Parmenides and Zeno, both of Elea.

We can tell these men few concepts that they have not thought of or heard before, but news of the great discoveries of modern science delights them. Parmenides is pleased to hear of the goal of our expedition, to unify the fields, for, he says, 'Truly, all things are one, and plurality is an illusion.' Zeno is skeptical, however, when we speak of matter being made basically of negatively charged electrons, protons of positive charge, and neutrons which have no charge. 'You sound like Democritus!' he growls, "but at least you admit that the atoms can be divided.' He offers his interpretation of space and time by telling us that Achilles cannot beat a tortoise in a foot race if the tortoise is given a head start, because by the time Achilles has halved the distance to the tortoise he must then travel half of the remaining distance and so on in an endless fractal relationship, so he shall never pass the tortoise. One of our party who likes foot races disagrees and offers to test Zeno's argument with an experiment, if a tortoise can be found for sale nearby in the market. We sweeten this offer further with a wager. Zeno laughs at this and walks away, saying we have misunderstood his whole argument.

Beyond Einstein: GEM Unification

Parmenides.

An annoying young character with them, named Socrates, stays behind and asks many questions of us, such as 'why are you climbing this mount Einstein?' We respond that we seek to unify the fields. He asks, 'why do seek you to unify gravity and electromagnetism?' 'So we can build a spaceship,' we say. He then smiles and asks, 'why do you desire a spaceship, since we live on a perfectly good planet already?' One of our party, of bolder makeup than the rest, finally says 'for the same reason you enlisted in the army to fight the Spartans, and went barefoot in the snow.' Socrates smiles at this and says, 'that is a good answer,' and walks away after his companions from Elea.

Beyond Einstein: GEM Unification

Zeno.

We encounter Eratosthenes, who having journeyed to Egypt, the wellspring of Greek thought, discovered a well at Syene on the Nile, where the Sun showed straight to the very bottom on one day a year. On this same day in Alexandria where Nile meets the sea, he measured the shadow cast by a stone column. From the angle of the shadow and its lack in Syene on the same day, he deduced the measure of the whole circle of Earth. It was of astonishing size compared to all the world he knew of or had heard of.

Beyond Einstein: GEM Unification

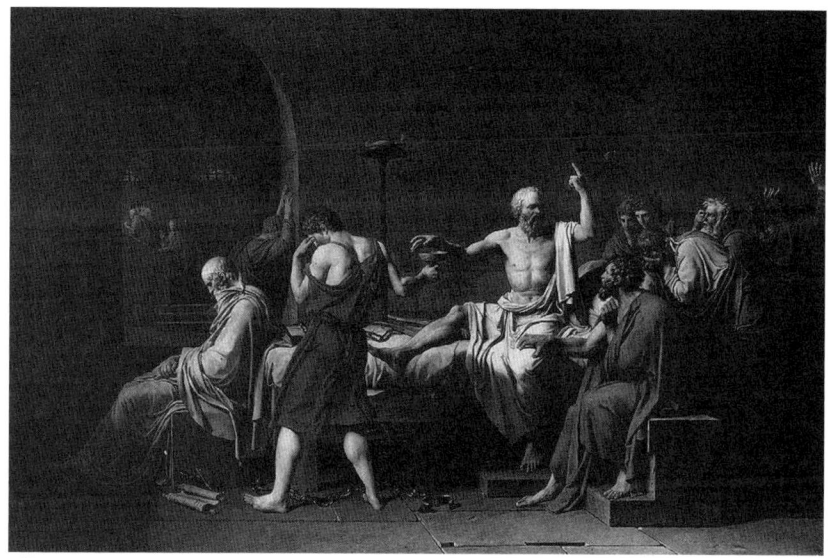

Socrates takes the Hemlock, preparing to hear the answer to his final question.

We further encounter Aristotle walking with Thales. They are both delighted to hear of the use of electricity and magnetism in the modern world, because they both wrote of these things in ancient times. We mention that Maxwell unified electricity and magnetism and they are puzzled, for they thought they were the same already. We mention the uses of the four states of matter, solid, liquid, gaseous, and the plasma state, and how we use these things to sail the vacuum of the heavens. Aristotle is pleased and comments that these are nothing but the five elements he spoke of in his works: earth, water, air, and fire, with the fifth element aether. He perceives gravity in terms of his five elements as a 'tendency for all elements to seek their place' as in stones seeking to rest on the earth, air bubbles seeking to rise, to join with the air, and fire seeking to rise above the air and join with the stars.

Beyond Einstein: GEM Unification

Ancient Athens, whose patron was Athena the goddess of wisdom.

To understand the cosmos, and the unities it contains, we must observe it as it is. The cosmos is expanding, uniform, and its mass is three quarters hydrogen and the rest predominantly helium. The helium is the natural product of hydrogen when it is hot and dense. This is how the stars make their energy. So this one quarter of helium suggests that the cosmos was once much smaller and denser, all one star, and as if it exploded from one point and was once made of pure hydrogen. The movement of the faraway galaxies reinforces this idea for they seem to be moving away from us in every direction, with those farther away moving faster from us as if at one time they all exploded from a compact mass. Based on the motions of the galaxies, this great cosmic explosion occurred approximately 14 billion years ago. This age is called the Hubble Time. That is three times the age of Earth. The galaxies are around us like snowflakes in a blizzard, fading off to a misty blue glow in the distance. They are as specks of dust.

Aristotle, the unifier of knowledge in ancient times.

Our galaxy, which appears typical among other galaxies, is like a spot on a spotted balloon that is inflating, and thus we are moving away from all the other spots. If we look 14 billion light years away, the spots are moving away from us at nearly the speed of light. These galaxies are far away, their light is dimmed because of their great speed away from us so we can barely see them. If they lie beyond, a bit farther out, we cannot see them at all. This radius is called the Hubble radius, and is the distance the light from the great explosion has traveled since creation occurred. There is a peculiar thing about this great creation explosion, which seems like it would be utterly chaotic: It was not disordered. The cosmos instead satisfies a delicate balance condition called "flatness" as if the explosion was precisely calibrated in its effects.

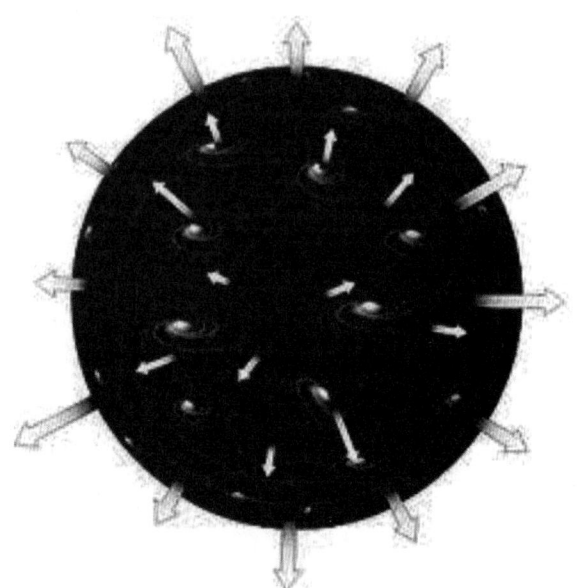

The expanding universe (NASA).

Beyond Einstein: GEM Unification

Our very own flying saucer: our view from inside the disk of our galaxy, the Milky Way.

A spiral galaxy similar to our own.

To understand what is meant by the flatness of the cosmos, one must understand that the cosmos is inherently dynamic and cannot be still. The cosmos is composed of empty space and particles of mass that all attract each other

Beyond Einstein: GEM Unification

with gravity. If the universe began as a scattering of particles all at rest, they would immediately begin to collapse on themselves because every particle would attract every other one. The universe would collapse to a point of infinite density, in reverse of the explosion that created it. The only way to avoid this collapse into a point is for the universe to expand so that every particle moves away from every other particle fast enough to overcome their mutual gravity. If the particles move away from each other too slowly, the universe will expand, stop, and collapse, just as a car with a stalled engine cannot make it up a steep hill. It is like a perfectly round pearl released to move in the bottom of a bowl: if the pearl moves too fast it flies out of the bowl forever; if it moves too slowly it is trapped near the bottom. However, if it is released at just the right speed, it will climb to perch on the lip of the bowl exactly, and that is how the cosmos looks. It looks like it was precisely tuned to balance the atoms and the void in contains.

Hubble deep field, with galaxies like snowflakes in a snow storm.

Beyond Einstein: GEM Unification

If the particles move away from each other too fast the universe will expand and not slow due to gravity, acting almost as if there is no gravity. Its bonds are broken. However, if the particles move away from each other at just the right rate, then the universe slows due to gravity but does not stop. It expands forever but the particles all interact by gravity and slow appreciably. It is perfectly balanced, so to speak, and gravity and expansion are both equal players in its dynamics. Cosmologists refer to this condition technically with the symbol Ω, and the condition of flatness of the universe as Ω equals unity, or one. It is appropriate that Ω, the last letter of the Greek alphabet, is used because its value determines the end of the cosmos. For Ω being near zero means the cosmos expands wildly and gravity plays little role in its dynamics. This cosmos ends in ice, for this cosmos expands and cools so rapidly few stars or galaxies can form in it. If Ω is much greater, then the cosmos ends in fire, since the cosmos first expands and then collapses back on itself in a white hot stagnation. But when Ω equals one, the universe balances expansion and gravity, and gravity locally can form galaxies and stars, so it expands endlessly and is a cheerful place while doing so, full of light and life. This balance is called the condition of "flatness," like the flatness of balance beam at equilibrium. This flatness made the creation of the universe an irrevocable act; it occurred once, it will not occur again, forever.

This condition is called flatness for another reason that will be discussed in more detail later. Briefly, Einstein discovered that spacetime is curved, and that the curvature of the pure vacuum and the curvature induced by the presence of mass-energy cancel at this balance point. So spacetime itself is flat.

Beyond Einstein: GEM Unification

This condition that appears to prevail in the cosmos, a situation of perfect and unlikely balance, seems, at first, to defy all reason. It is as if the eruption of a volcano had left a stack of volcanic glass champagne glasses, one on top of the other, stacked 1000 yards tall, and perfectly balanced. It requires the density of matter and the rate of the universe's expansion to be exactly right. However, in an unstable universe, the balance must collapse due to gravity or expand violently as if gravity did not exist; the condition of flatness means that the universe expands yet does so gently, and that gravity between the particles of the universe will overwhelm expansion locally, and lead to galaxies full of stars. This puzzle of flatness will be discussed again later on our journey, but now we simply accept it like the light of the stars at night.

To be at home in a cosmos full of mystery, one must rest occasionally on the truth that the cosmos is basically a good place to live, providing all that is required for life, and rejoice in this. The cosmos presents puzzles that can be separated and solved in a succession of approximations. To pluck a flower is to disturb a star, but one can ignore the stars while gathering wild flowers for one's sweetheart, to good approximation. If it was otherwise, the cosmos would be one huge tangle of mystery; instead, it is a succession of smaller mysteries. But mysteries have no end, and it is beyond flesh and blood to understand them all, so we must rest content after solving one after a period of wrestling with it. After resting in this thought, we again renew our journey.

The stars shed light at night and the Sun gives warmth in the day because the universe is made mostly of hydrogen;, that is, a proton, which is heavy and has positive electric charge and an electron, which is low in mass and charged negatively. The electric force binds the oppositely charged

Beyond Einstein: GEM Unification

proton and electron together to form neutral hydrogen. The electron orbits the proton like a planet orbits the sun. The gravity force then gathers the hydrogen atoms into collapsing clouds that form stars and when the heat of contraction and the density in the heart of the stars reaches a million degrees the stars unleash a new energy source called fusion. In fusion the hydrogen fuses to form helium and releases energy. This power, which lights and heats the universe, comes from the strong force. However, more subtleties are involved.

Beyond Einstein: GEM Unification

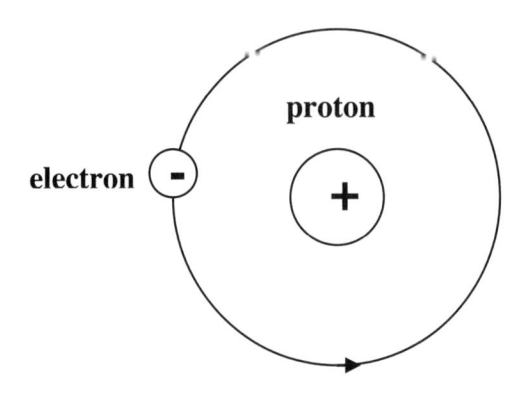

The hydrogen atom, the basic building block of the cosmos

The universe, despite all the variety of elements that it comprises, at least 100 that are stable, is made from only three basic particles: the proton, the electron, and the neutron. Put every strange particle produced in every particle collider in a box and look in the box 13 minutes later and you will find only hydrogen, protons, and electrons. The neutron is unstable by itself and decays into an electron and proton plus a tiny wisp of a particle called a neutrino, which has no charge and may have no mass, that whizzes off at near the speed of light from this decay. No box can hold it, it passes through planets without stopping. This is how the weak force makes its presence known. Without the neutrino, which barely exists, the neutron is basically just highly compressed hydrogen. However, there is a deep riddle here:

Beyond Einstein: GEM Unification

The neutrino carries spin, a spin as mighty as any electron or proton.

Particle	Proton	Neutron	Electron
Charge	+	0	−
Lifetime	Stable	890 sec	Stable

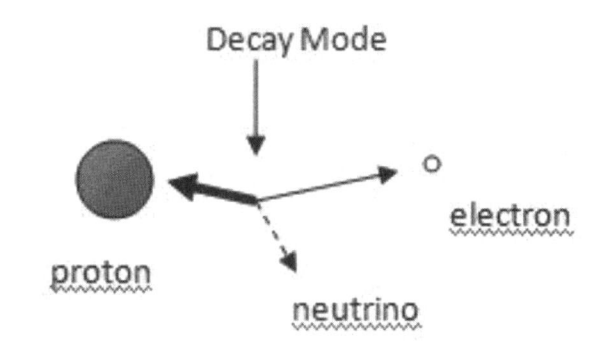

Decay Mode

proton / electron / neutrino

Long lived fundamental particles.

Spin is a property found everywhere we look in the cosmos. The galaxies spin, the stars and the Sun all spin about their axes like gyroscopes, the planets all spin likewise. So it is when we look into the atoms and subatomic world, all the electrons and protons spin. They spin with the amount of spin predicted by Planck's fundamental quantum of action that requires that nothing, not even the vacuum itself, can rest quietly. By the laws of quantum mechanics all

Beyond Einstein: GEM Unification

five elements, including the aether itself, must be full of activity and vibration. The basic particles of the cosmos spin so fast they act as if they spin faster than light, if that was possible. They spin like spinning ice skaters on a perfect ice rink, never stopping, never slowing down. The action of the vacuum is called the ZPF or Zero Point Fluctuation.

The three basic particles, protons, electrons, and neutrons, are therefore, in a basic sense, only two, the proton and electron, and they can explain the basic universe as it is. The proton, by its great mass generates the gravity of the universe, and the electron because of nimbleness generates all sensible electromagnetic phenomena. These two basic particles form a pair: the proton and electron and two basic forces, gravity and electromagnetism that arise from their dynamics. The proton, because of its weight creates gravity; the electron, because of its nimbleness, generates light. Around the proton moves the electron, to form hydrogen, and around the stars formed of hydrogen, move planets, as the electrons move around the protons.

Whenever the electron moves, it generates electric and magnetic fields, these form the magic elixir that is light. It is like moving your fingertip on water, creating ripples. By this light created by moving electrons, the human race sees the grand tableau that is the cosmos, with its galaxies, stars, and planets. Thus, our basic thumbnail sketch of the universe is complete. It is grand, and full of stars and nebulous light; its great dynamics arise because of gravity and this dynamics is seen in electromagnetism. The cosmos contains two basic particles, protons and electrons, and two basic forces, gravity and electromagnetism. It is full of light, and expands at just the right rate to give gravity an equal role in its dynamics.

Here is a great riddle of the universe as it is, discovered by a genius named Paul Dirac. Because both gravity and electromagnetism follow the same force law variation,

Beyond Einstein: GEM Unification

decreasing with the distance squared, the force of electromagnetism between an electron and a proton and the force of gravity have always the same ratio, no matter if they are separated by the radius of the visible universe, which is the Hubble radius. Such is the correspondence of their dynamics that if one force disappeared the electron and proton would continue their dance, only changing the size of their ballroom.

Dirac had become famous as a physicist in 1928 by discovering that each type of matter had a corresponding mirror image called antimatter. Dirac was an outspoken atheist, so much so that at one scientific conference, Wolfgang Pauli proclaimed *" There is no God, and Dirac is His Prophet!"* Pauli would win the Nobel Prize for his discovery of the "Pauli Exclusion Principle" that required that electrons with the same spin repel each other, causing matter to occupy space.

Dirac had made his discovery of antimatter by studying the riddle of particle spin, the fact that the electrons and protons, and even the bashful neutrino, had spin, as if they were tiny gyroscopes. Ironically, accommodating this important detail had led to an epic discovery. The antimatter particles had an opposite electric charge, and the same mass as their normal matter counterparts, and when they encountered their normal counterparts they both instantly flashed into electromagnetic energy in the form of gamma rays. They annihilate each other, leaving nothing but vacuum and deadly radiation. The anti-electron was called the positron, because of its positive charge, whereas the antiproton, a much rarer and harder to create particle, was merely called by that name. It was the ancient idea of the nemesis or "doppelganger" returned. Dirac postulated that what he perceived as empty vacuum was in fact a "sea" of particles and antiparticles appearing, moving slightly apart

Beyond Einstein: GEM Unification

then collapsing back into electromagnetic energy of the ZPF. An antiparticle-particle pair is effectively an excited state of the vacuum. Their charges add to zero and their masses are identical, so it is as if their properties cancel. From the vacuum these particles arose, and to the vacuum they will return. This situation is completely different from the case of the electron and proton, which, despite their opposite charges, have such different properties that they cannot cancel. Physicists have speculated that the universe began from the vacuum as matter and antimatter, arising as it were from the "Dirac sea," but this would not result in the universe we experience; it would be a doomed place full of gamma rays and unable to sustain life.

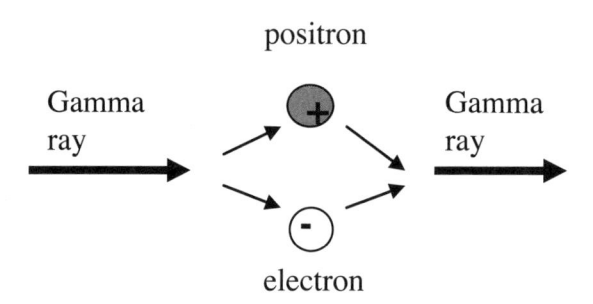

Particle-antiparticle appearance from the vacuum ZPF: The pair appears from a ZPF gamma ray, then recombines and annihilates to reform the gamma ray in accordance with Heisenberg's Uncertainty Principle.

Instead of the "Dirac sea," it appears that the protons and electrons making up the universe at its moment of creation arose from some other sea. This "Primal Sea" was a vacuum, like the Dirac sea, and charge neutral, but from it sprang particles, electrons and protons, whose properties did not

Beyond Einstein: GEM Unification

cancel exactly, like matter and antimatter, but instead became the stuff of existence. Like the answer to any riddle in the cosmos, the riddle of particle spin led to a deeper riddle, the riddle of existence.

Paul Dirac, who linked the cosmic and the subatomic.

A decade after his discovery of how to understand spin, an inherent property of particles, and the resulting discovery of antimatter, Dirac was emboldened to measure the temple of the universe. For a measuring stick he chose not any human measurement, a human foot or arm's length. He instead chose the width of an electron.

Beyond Einstein: GEM Unification

Wolfgang Pauli, the father of exclusion.

Because of Einstein, Dirac knew that all mass is energy. Indeed, it is this energy which appears when matter and antimatter come together and annihilate each other. He also knew that the electromagnetic force stores energy in space. Charge a silver sphere with electricity, shrink it, and it will weigh more. It will do this because moving the charges together on the sphere's surface when you shrink it will take energy, because the charges repel. Take a sphere of pure charge weighing nothing, shrink it enough, and it will acquire the mass of an electron. The radius of the sphere when this occurs is called the "classical electron radius" or simply electron radius for short. Nor is this a simple thought experiment: shine a laser though a vacuum with one electron and the electron scatters light as if the vacuum was a silver sphere of this radius. So Dirac found a fundamental unit of length, determined by the cosmos when it formed. For his unit of time he discarded the human heartbeat, the origin of the second, and used instead the time it took the speed of light to cross this electron radius. For his unit of weight he chose not a pint of ale, but the electron mass. Thus armed

Beyond Einstein: GEM Unification

with a bag of cosmic weights and measures, he set out to measure the universe. His measurements used a logarithmic scale, which is to say, orders of magnitude. Using logarithmic numbers one can compare the very great and the very small number of the universe on the same piece of paper. What Dirac discovered was as astonishing as when Eratosthenes first measured the size of the Earth.

When Dirac measured everything with his new set of weights and measures he found that everything that physics cared about was approximately unity. The size of the atoms, the weights of the particles, the energies and motions, all came out as numbers within an order of magnitude or two of one. But two numbers in the cosmos stuck out like sore thumbs.

One number was the ratio of the strength of EM to gravity forces between electrons. This golden ratio was an enormous number of approximately ten to the 42nd power. The other number was equally astonishing; it was the radius of the cosmos, the Hubble radius, measured in terms of electron radii. It was literally the largest distance measured by the width of the smallest subatomic particle. What was even more astonishing was that it was almost the same number: 10 to the 42nd power. The two mightiest numbers in the cosmos, the ratio of EM force to gravity between a pair of the two most common subatomic particles in the universe, and the size of the cosmos measured in subatomic dimensions, were almost exactly the same. It was as if two twins separated at birth and living in different states all their lives had both won the National Powerball lottery in the same drawing.

But how could this be true? Dirac speculated that because the radius of the universe must grow large in time that the ratio of EM to gravity forces must be growing larger. But repeated checks of this, the most definitive by one of my

Beyond Einstein: GEM Unification

former college professors, Ronald Hellings, have shown that the ratio of forces has remained constant. So it is a fine riddle of the cosmos.

This riddle can be told another way: if there was no EM force, and we could not see the glow of the first moments of creation at the Hubble radius, would gravity, which determines this radius, exist?

It is seen that the Hubble radius is two things, it is the distance light has traveled since the dawn of creation, but it is also nearly the distance that light travels before it scatters off an electron in a flat universe. The universe is slightly misty, not clear as crystal; it is misty because of the matter that resides in it, and the mist is of electrons which scatter light. Because the universe is gravitationally flat, and is also slightly misty, the two balances are nearly the same, and that is the answer to Dirac's riddle (this is shown mathematically in the chapter note.) This is another clue that the gravity and EM forces are paired, like twins, yet not identical twins, but still twins separated at birth.

Thus the cosmos consists of nothing but atoms and the void, but the atoms are rich and complex, composed of diverse particles, born basically from electrons and protons, bound together by two forces of rich character, EM and gravity. The void itself hums with quantum activity and its vastness is related to the ratios of its smallest parts. It is expanding in perfect balance.

Chapter 2. A Book of Numbers and Forms

"The most incomprehensible thing about the universe is that it is comprehensible."

Albert Einstein

"All of life is twilight moving into a dawn ... ."

Edwin Wright

"All is Number"

Pythagoras

We journey north from Athens, and stop at the shrine to Pythagoras and speak to the Pythagoreans, his disciples. They believe all the universe can be understood by numbers, after the teaching of their founder. We disclose to them our journey's purpose: to climb to the summit of Mount Einstein, where, it is said, the stars can be seen at noonday. They are pleased and disclose some of their mysteries to us, in the

Beyond Einstein: GEM Unification

hope that this will aid our journey. They warn us that the important numbers of the cosmos are not all perfect as they had supposed.

All perfect numbers are ratios. These numbers are called perfectly rational. However, some numbers, they warn, are irrational, and make no sense, and cannot be rendered perfectly. They can be found only by successive approximation, where a number is rendered in stages to perfection, though this stage can never be reached. One such number is π, which squares the circle, and is useful in all things yet cannot be discovered as a perfect ratio, only as an approximation. This truth caused them great pain, as it did their founder. Another number is the diagonal of a perfect square, which is also irrational. They suggest that it is the squares of numbers that are important, not their square roots.

'How is it that that which is perfect can give rise to that which is imperfect?' One of the Pythagoreans laments. One of us volunteers, 'the cosmos and all it contains is imperfect, yet it arose from simple perfection. Perhaps imperfection must exist, if only so we children of the dust may exist?'

'The vacuum is perfect,' one volunteers, 'yet it is sterile, so matter, whose forms are imperfect, must arise to allow existence.' We leave them in a great debate about this.

Another group discusses the teaching of Pythagoras who said that all the universe runs on numbers, like the tones of a plucked string progressing through an octave in a harmonic series. Each element is like a note or chord, together they make up the symphony of life.

Beyond Einstein: GEM Unification

Pythagoras, the first philosopher of numbers.

They also point out the five perfect solid shapes that can exist in our three- dimensional space. There are five. Three have regular triangles as faces: the tetrahedron, octahedron, and icosahedron. One exists with square faces. Finally and most important of all, one exists with twelve regular pentagons as faces. They regard this fifth regular solid, the dodecahedron, as the last and most complex and significant of all, modeling all the cosmos. The pentagon embodies the golden ratio of 1.628 ... which is fractal and from which, they say, the whole measure of the universe from its largest systems to its smallest can be understood.

We tell them we wish to unify the two great forces from the four in nature. They insist a fifth force must exist because there are five perfect solids. We suggest that there are only four dimensions of spacetime: height, width, depth, and time, so why should there not be four forces. They respond

Beyond Einstein: GEM Unification

and say there must be, therefore, five dimensions, one for each perfect solid, and that we are obviously miscounting. We are chagrined and go on our way.

We must have special preparation for our journey here in the shrine of numbers. To make it to the summit of mount Einstein special mountaineering equipment is required and special training is necessary. The climb is not arduous if one follows the right path, but still it requires discipline and new skills not required for a walk around the park. You must learn to think in terms of important numbers and forms. This will not be hard; you can count the number of shapes on your fingers. However, you must learn also to use both real and imaginary fingers to reach this summit and grab handholds with them. You must measure distance in space and time. So prepared, we can attain great heights, and finally stand in a place where heaven and earth are unified and the stars are visible at noonday. Let us prepare for the next stages of the journey.

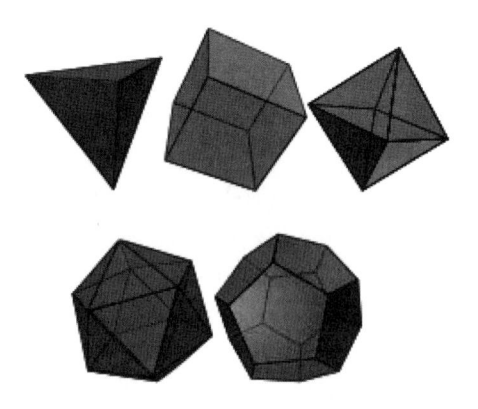

The Five Platonic solids.

Beyond Einstein: GEM Unification

We judge there are four dimensions because we can move an object in one direction but measurement of the object's position in the other three dimensions seems unchanged. That is, an object raised above the floor in a room does not move from its initial spot on the floor in terms of depth or width of the room. We label the dimensions x, y, z, and t, for width, depth, height, and time. An object in its place will not change its position in space, in our room, even as time passes. This is the reality we are born into and for which our senses and thought processes have been optimized.

"Behold," **the caption of this ancient diagram proving the Pythagorean theorem in one image.**

However, when we do such things we are also naming a fifth dimension by implication. We can define a dimension as a viewpoint by which an object can be moved so its position does not change. The dimensions are, therefore, called orthogonal, meaning "true," because they are independent of other dimensions. So it is with height, depth, width, and time. They are orthogonal and do change when the others are changed. However, this means a fifth dimension must exist that is also orthogonal. This is the dimension by which we identify the object itself, for if we move an object its identity remains unchanged. Its

Beyond Einstein: GEM Unification

changelessness itself, the fact that it does not change its identity, defines an orthogonal fifth dimension. The fifth dimension is the dimension of identity. Let us call the dimension "objectiveness" for now. We must name it to measure what does not change when we change the other dimensions. We can reduce this to a simple case. Our object is a red ball.

The red ball has two attributes, it is a ball of a certain size and it is red. If we move it up, down, sideways, back and forth, it remains red and it remains a ball of the same size. By this we establish that there are things that are independent of space. We let the ball sit for a while; it does not change, it remains red and it remains a ball. The ball possesses dimensions that are independent of time. The ball consists of a set of points, a center, and the points making up its surface. We can rotate the ball and it looks the same from all sides. This is a property of the spacetime in which we live. The Pythagorean theorem, that the square of the hypotenuse of the triangle is the sum of the squares of the sides, makes this so. The spherical character is preserved because space is isotropic and flat, not curved. The distances between the center and surface points are always the same, no matter how we turn the ball.

However, the redness of the ball is an attribute that has nothing to do with space, it is instead that property of how the wavelengths of electromagnetic radiation interact with the surface. This is determined by, in space, a tiny layer of molecules which process light so as to only reflect red light and absorb all others.

So a sphere can be conceived as a collection of points independent of how it moves and of time. Yet it must be seen, and its appearance itself comprises a fifth dimension. Dimensionality is a collective truth, like a word that has no

Beyond Einstein: GEM Unification

definition except in the context of other words, in a language.

A proton is an object with an identity that does not change when it moves in space and time. The same can be said of an electron. So we can define "electron-ness" and "proton-ness" as dimensions. Because they are always a pair and close to each other, we will say that together the electron and proton define an extended object like the red ball of our previous discussion. They exist as two ends of an extended object in an abstract fifth dimension, one being the plus end and the other being the positive end. Remember this.

Numbers are the language in which many of the truths of the cosmos are written. Numbers begin with our five fingers, and if necessity required, our toes. Zero, representing the void, was late in coming and was invented by the Arabs. The Mayans also independently discovered it. We count by tens because we have five fingers and two hands. The Chaldeans counted by 60s, which significance is lost to us, but could represent a flock of sheep. It is five dozen. Because of the Chaldeans' reverence for the number 60 we divide the circle into 360 degrees, and the minute and hour likewise into 60 parts.

Beyond Einstein: GEM Unification

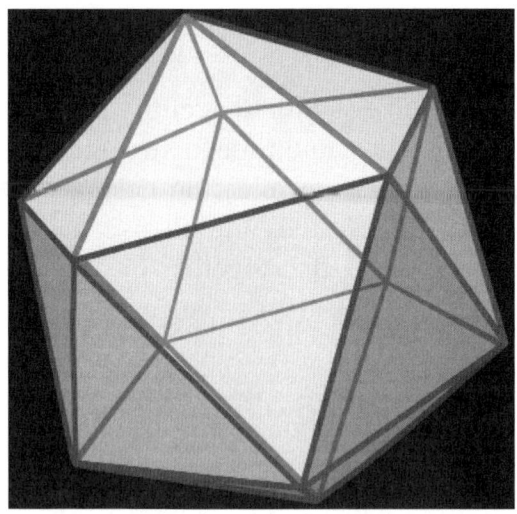

An icosahedron.

The counting numbers we call integers, for they count things that have integrity; that is, they represent things that are independent and exist by themselves. It was discovered that integers can be negative, as in one "of our sheep is missing." The whole numbers are the integers plus zero, the void, which expresses the number of sheep missing when the missing one is not found.

The whole numbers can be laid out on a line called the number line. The spaces between the whole numbers can be filled with fractional numbers. The numbers are ratios of whole numbers to others. As in "I will divide my stack of hay into 60 parts for my flock of sheep so each will feed equally." These numbers are called rational. However, they cannot describe the cosmos as it is. Other numbers are required to fill the spaces in the number line. These numbers, like π, or the square root of two, are irrational, they can only be approximated to useful accuracy, like life itself. These classes of numbers taken together are called the real numbers, because they correspond to reality as we

Beyond Einstein: GEM Unification

experience it everyday. But we spend one quarter of our life dreaming, either while asleep or waking. What numbers describe our dream state?

Some people believe that all truths are scientific; that is, they can be dissected on a table, or seen clearly in a test tube. This is to say, we can only understand what we can control. But some truths are beyond our control. Some truths cannot be seen in a test tube, only in the cosmos itself. Likewise, some things cannot be captured and dissected; they must be studied alive and free, in their own place, with the knowledge that to observe them closely is to perturb them.

The existence of negative numbers is an abstraction in everyday life. That is, unless your checking account gets overdrawn. In the larger reality of the cosmos negative numbers are as real as any star or atom. In the study of electricity, negative charge exists as surely as positive, and can give just as stunning a shock. Protons are positive and electrons are negative. Antimatter can be understood as a negative number, the mirror image of ordinary matter. Encounter your mirror image out in space and you will annihilate each other.

Likewise the square root is a real number. To take the square root of a number is to solve for the hypotenuse of a triangle. Draw a square with unit sides, then draw its diagonal and the Pythagorean theorem gives you, behold, an irrational number. Together the irrational and rational numbers make up the real number line. Discovery of this truth caused Pythagoras himself great consternation, and kept it as a great secret of his society. Now we take another bold step beyond Pythagoras into the unseen. We take the square root of a number which is negative. This number is not on the real number line; it is imaginary. However, we will find that things that can only be described by imaginary

Beyond Einstein: GEM Unification

numbers have profound consequences in reality. In the cosmos as a whole, we will find the truism of Einstein, that

"imagination is more important than knowledge."

Many numbers we deal with in physics are neither completely real or completely imaginary, they are a combination of both. These numbers are complex. It can thus be said with mathematical precision, that the complex universe we live in runs on quantities that are partly real and partly imaginary.

The cosmos is flat to good approximation. Light follows approximately straight lines everywhere; if it curves, it is usually slightly and even then it continues straight as before. The cosmos appears balanced between atoms and void. We see this because galaxies farther away, judged by their starlight, look smaller in the same proportion. Spacetime is like a flat sheet of paper. One point defines a number on the number line. Two points define a distance between them. If we draw three points on a sheet of paper and draw straight lines between them we make a triangle, the most basic closed shape. Make the points equidistant and the triangle has interior angles of 60 degrees, pleasing to the Chaldeans, and divides the circle into three equal parts. The three points also define a plane, like a flat sheet of paper.

There are only five platonic solids; that is, five closed shapes in three dimensions whose faces are regular polygons, having sides of identical lengths. They are beautiful to behold and study. The simplest is the tetrahedron, with four faces each of a triangle of equal sides. The next shape with triangular faces is the octahedron or eight sides, this shape looks like the great pyramid of the Egyptians viewed with its reflection in the Nilebut is two identical pyramids joined at their bases. This joining forms a

Beyond Einstein: GEM Unification

square. We can do more with the triangular face. We can construct an icosahedron based on joining the triangles in groups of five to form domed pentagonal shapes.

If we take the square as the basic polygon, we have the favorite shape of all measures of pastures and floors. The square is the fundamental shape demonstrating two independent dimensions that are merely rotations of one another in an isotropic universe whose underlying shape is a circle or sphere. The platonic solid we construct of squares is the cube, which is the measurer of spaces. This is the fourth platonic solid.

The fifth and most profound of the solids has twelve faces, the dodecahedron, the simplest of the shapes with the most complex faces. It has twelve faces of five sides each, pentagons. The pentagon is a shape that is based on the golden ratio of φ which is 1.628...... The Pythagoreans were convinced that the dodecahedron profoundly related to the Cosmos, and represented the aether. It is the fifth regular solid, giving five fundamental dimensions necessary to understand the basic cosmos.

Beyond Einstein: GEM Unification

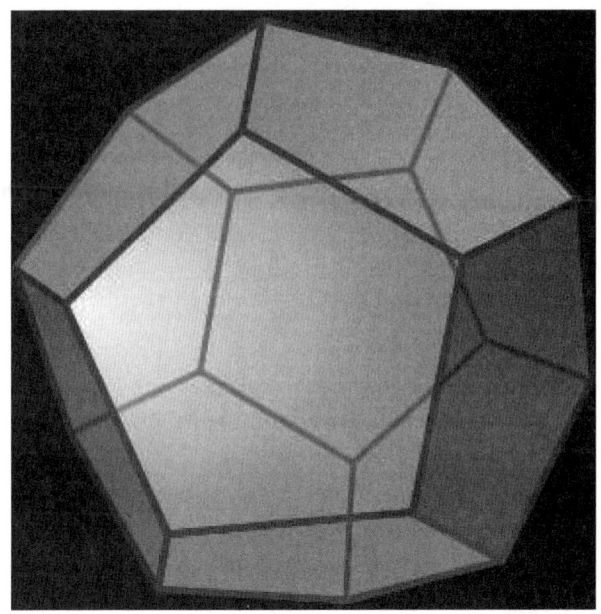

A dodecahedron: representing the aether.

The golden section carries in it objects with similitude at all scales, which is called in modern terms a fractal. It is the golden rectangle that can be subdivided endlessly into more golden rectangles. These patterns look the same on the tabletop as they do under a microscope.

The simplest shape of all cannot be drawn on paper. It is the sphere, which the Greeks believed was the perfect shape. It embodies the circle, since we see images on our retinas that are two dimensional, and presently looks like the same circle when viewed from any side. It says the dimensions are all equal, and the universe is flat. Just as the dodecahedron is the embodiment of φ so the sphere is the embodiment of π. The number φ is related closely to five and speaks of the poetic beauty of the universe. The number, π, we shall see, speaks to the brilliance and power of the cosmos.

Beyond Einstein: GEM Unification

The sphere is the embodiment of power and dynamics in the cosmos. The cosmos is a sphere looking the same wherever one looks. Earth is a sphere, the Moon likewise. The Sun and stars are spheres. The atoms and the proton and the electron are spheres.

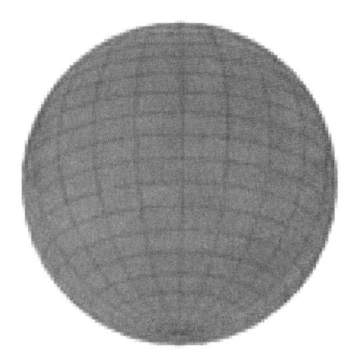

Sphere.

The sphere exists in four-dimensional spacetime as an object independent of time as it is of space, but what of its counterpart in four-dimensional space? Our minds move along the axis of time, so we see reality only as a section or slice. The past is but memory, the future but a dream. We cannot see a spacetime object except as a movie, a dream. A dream of a sphere is merely a sphere changeless, but what of a true sphere in four space?

A hyper-sphere, a 4-d sphere, a sphere with 4 spatial dimensions, can only be seen as a shadow or slice in three-space. By analogy a 3-d sphere can be seen by a two dimensional being as a series of circles, or perhaps a set of concentric circles, each 2-d spheres, one inside the other. So we can see a hyper-sphere, a 4-d sphere, as a sphere full of concentric spheres. From distance it appears as a sphere with internal structure.

Beyond Einstein: GEM Unification

Einstein discovered, in his Special theory of Relativity, a great mystery, which speaks to our limited Earthly view. The truths of the whole cosmos are partly Earthly and partly unearthly. The truth of distance in spacetime is unearthly. Unlike the other dimensions whose squares add, according to Pythagoras, the square of the time dimension must be subtracted from the squares of the spatial distances. That is the way it all works. This means time is a different sort of dimension than height, depth, and width. Some have said that the difference of time from space is an illusion, but its difference is real, according to Einstein. Space has three subdimensions, space is triune, but time is unitary, and between them is a negative sign.

Special Relativity means the four-sphere is not merely a hyper-sphere, an object with an additional fourth special dimension, but a dynamic entity. The negative sign means a four-sphere cannot sit still; it must expand or implode as we watch it from our perch in three space. One sphere expands as we move endlessly from the past to the future. The other four-sphere implodes in our sight as it moves from the future to the past. What we imagine as a sphere in three space independent of time, is in fact the collision of the two four-spheres as they move through each other, one expanding from the past, the other imploding from the future. So all things we regard as permanent are actually the past and future colliding in front of us. It is the twilight of yesterday unified with the dawn of tomorrow. The past is a memory and the future is a dream; where they meet is this moment.

The next shape embodying π is the torus, the shape of a wedding ring or donut. It is the shape that embodies endlessness and spin. It is the shape that embodies another great shape of power in the cosmos, that of the vortex, the smoke ring. The vortex is generated wherever the cosmos concentrates great power. The line of the core of the vortex

Beyond Einstein: GEM Unification

cannot end in the fluid in which it forms; it must rest at the fluid's end, as in a tornado ending on the ground and the clouds or else continuing in an endless loop, partly unseen throughout the cosmos. The shape of a vortex then, in an endless universe with no boundaries in space, is the torus. Another vortex shape is the place of vortex concentration that can be seen, and the diffuse vortex lines in space that reconnect through the vortex.

Torus.

In the great classic, *The Book of the Five Rings,* the great warrior Mushashi wrote of the five elements, earth, water, wind, fire, and the void.

Beyond Einstein: GEM Unification

Miyamoto Musashi, the quintessential samurai.

Take a bowl and fill it with water. Touch its surface and waves race in circles from your finger tips. To move one finger is to make one circle of ripples, to move two is to create a pattern of expanding ripples. The waves, when they are small, move through one another without affecting each other. When they pass each other they create a pattern of interference, in some places the peaks of the waves add, in others they subtract. Waves are the fabric of the cosmos. To analyze the waves one sees one must understand the geometric functions of the circle.

Beyond Einstein: GEM Unification

Ripples on Water.

Draw a circle, divide it into quarters by lines that cross at right angles at its center. Then draw a radius to its edge, draw a line from where the radius crosses the circle to make a right angle with one of the crossed lines in the center. This defines a right triangle. Now sweep the radial line around the circle and the functions it defines by the lengths of the base and height of the right triangle are the sine and cosine. They are sinuous functions describing waves. They are harmonic motions. Thus is the wave derived from the circle.

Beyond Einstein: GEM Unification

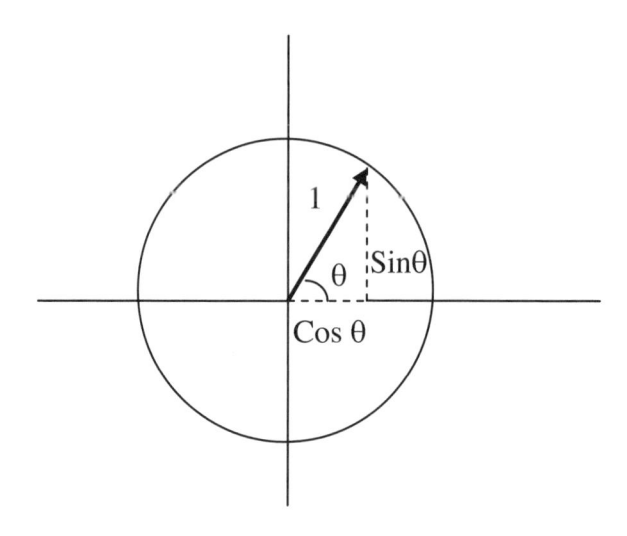

The functions Sine and Cosine as projections of an arrow on a circle of radius one.

Another form of nature is the form of growth. Take a single grain of wheat and plant it, then when it grows harvest the new seeds from its ear, plant them again, so after many seasons, the single grain will have filled a field. This pattern of growth, where the rate of growth is proportional to the growth itself, we will call exponential growth. Nature literally thrives on exponential growth. It uses an irrational number, "e," which is approximately 2.718... . The EXP function, the function of self-feeding growth, can be understood as a harmonic function of an imaginary geometric angle. It can be visualized by drawing the projection of an angle onto a curve called a hyperbola instead of a circle. The hyperbola represents the imaginary nature of the angles.

Beyond Einstein: GEM Unification

Instead of projections on a circle, the projections are made onto a "funhouse mirror" that magnifies.

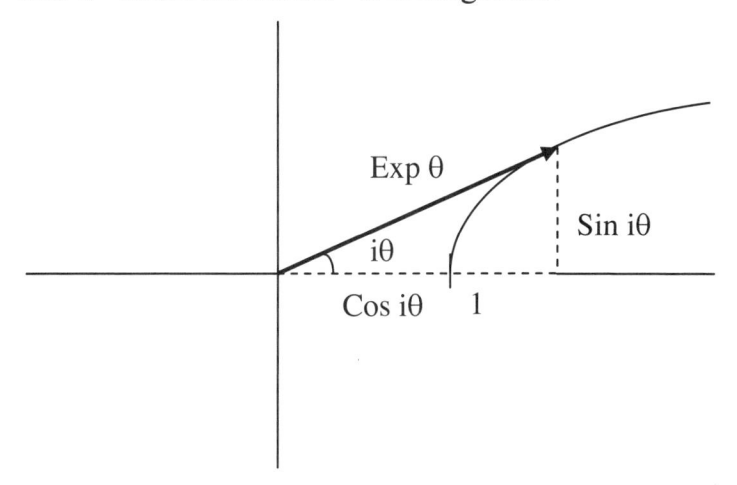

The exponential or Exp function as a length of an arrow projected on a hyperbolic curve of minimum radius one.

Using the description of numbers as exponents, one can multiply numbers by adding their logarithms, which are the exponents of the numbers, and then taking the exponential of the result. Exponential expressions allow one to write the measures of the cosmos and the atom on a piece of paper and compare them. Such relationships arise often in nature, where the small determines the fate of the large. So this is a very powerful way of expressing numbers.

Living things grow and expand exponentially in number, time, and space. This is the pattern of the living and of the ideas that animate the living. Behold, there is mystery here, for the wheat grains cannot multiply except they are pollinated. Some unseen thing in the air makes the wheat fertile. Like the tip of an iceberg, which owes its grandeur to a vast body of ice beneath the sea, so is the known compared to the mystery that enables it to exist and be studied.

In my youth I cared for the vast green lawn of my family's home. To make sure it was well groomed, well watered, lush and green was my responsibility, in which I took great pride. In caring for it through high school I learned every corner of it. One day I discovered that a circle of it had died and turned pale yellow. No matter what I tried the circle of grass could not be revived. However, I was blessed by a fine circle of mushrooms that had emerged in the dead grass the next year. The mystery was solved, and another mystery was revealed. I had always thought that mushrooms were isolated things, but now I knew they were part of a larger organism that lay hidden beneath the ground. So it is with the cosmos, that what we see separately is often the fingertips of a hidden hand.

I saw another marvelous thing growing up. I watched a magnificent summer whirlwind cross an empty field by a busy road. It was large and powerful, with a vortex of leaves and papers three yards across. As it crossed the road, a large and powerful semi-truck came barreling along and struck it. The truck passed through it without disturbance, blowing leaves and papers in every direction. After all, it was tons of metal and cargo moving like the wind, and the beautiful whirlwind was merely wind and dust. However, to my joyous amazement, the whirlwind reformed immediately behind the truck, reassembled its load of leaves and papers, and went on its way as if nothing had happened. I saw from this that I was mistaken in thinking of the whirlwind as merely the core of frenzied motion that I could see, but instead saw that the vortex had structure that was actually far broader and taller than I could see. The main part of the

Beyond Einstein: GEM Unification

vortex hid in the clear air, while I was merely looking at its intense core. So it is in life, that the unseen rules what is seen. So it is with the cosmos, where all that is seen is but the visible portion of a vast and complex mechanism, most of which escapes the human eye.

The scientist speaks in the language of mathematics and the poet speaks in the language of words. The truths of both sustain the world. The language is the voice rendered into words and words in the West are rendered into letters of the alphabet. We seek to unify gravity and electromagnetism or EM. The letters E and M are from Electron (amber-"shining thing") in the Greek and the word magnet from Magnesia, a region of Asia Minor where lodestones were found. Magnet comes from Magna or "great" in the Greek. To the poet's keen eye electricity means the "bright force" and magnetism means the "great force." Gravity comes from the Latin word gravitas or weight. The letter G originates from the process by which the Romans, by successive approximation, rendered Latin into words by an expanding alphabet. G was invented as a combination of the Latin letter C and the Greek letter Γ, both the third letters in the respective alphabets. G is the broken circle of C and square angle of Γ, combined into one letter. G is the circle squared.

Thus have we prepared ourselves to journey to the province of Gravitas, Magnus, and Electra.

Beyond Einstein: GEM Unification

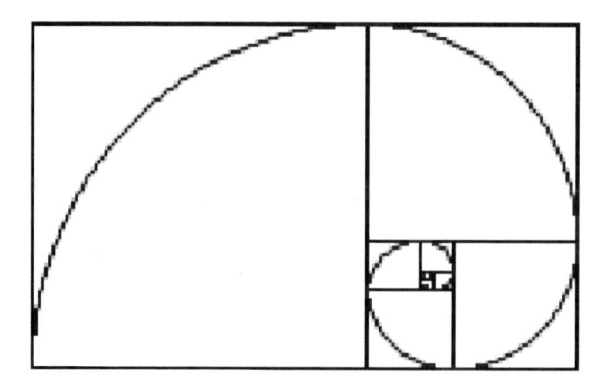

The Golden Spiral, which looks the same at all scales, demonstrating the fractal nature of reality.

Chapter 3. Kepler, Newton and the Sun King

"No great discovery was ever made without a bold guess"
Isaac Newton

From the diary of the expedition:

We journey now to Africa, to where Egypt adorns the Nile, and as Pythagoras did, we sit at the base of the pyramids and ask endless questions of the Sphinx, perhaps the oldest human sculpture. From the Pharaoh Akhenaton, we learn that the Sun is the one god and the center of the cosmos. We journey to Syene, and then to Alexandria, to take the measure of the Earth, and, using it as a golden ruler, to measure the heavens. Here we find the puzzle that led the wise to under stand the first force, gravity, and how it moves both Heaven and Earth.

The source of the Nile lies deep in Africa, the cradle of humanity. Where the Nile flows forms a rushing torrent across the desert to meet the Mediterranean, lies Egypt the source of Western Civilization. From Pharaoh Akhenaton came the first concept of the Sun as the center of existence. Egypt of the Pyramids and pharaohs was the teacher, and Israel and Greece sat at her feet. Everything Greece learned of science and mathematics it learned initially through Egypt, and after the Persians conquered Egypt the student excelled her teacher. In 332 BCE Alexander arrived to liberate Egypt from the Persians, and the grateful Egyptians declared him Pharaoh. He was determined to create a center of learning and trade in the world that he knew, and it must be done in Egypt, so he founded the city of Alexandria, with a great lighthouse, as a shining light to the whole world.

Aristotle documented that the Greeks knew the world was round in 100 BCE because it cast a circular shadow on the Moon during lunar eclipses. This disturbed no one because the sphere was a perfect shape to the Greeks.

Pharaoh Akhenaton as the Sphinx worshiping the Sun.

Beyond Einstein: GEM Unification

Aristarchus saw everything clearly from this shadow of Earth on the Moon. He was an expert geometer and mathematician, with a sharp mind unfettered by previous thoughts. He measured the Moon using the curve of Earth's shadow, and found it was roughly one-fourth the radius of the Earth. He then found from observing the angle made by the half Moon with the Sun that the Sun had to be a thousand times further away from us than the Moon. He knew from the eclipse of the Sun, that the Sun had to be then a thousand times larger than the Moon. Being Greek, the idea that the Sun, being much larger than the Earth, and most powerful thing in the Heavens, should revolve around Earth once a day seemed ridiculous. Rather, he saw that Earth and other planets revolved around the Sun, and that Moon, being smaller than the Earth, revolved around Earth. However, just as Akhenaton was opposed for promoting the Sun as the one God and center of existence, the followers of the lesser gods of Earth protested. Among the Greeks the complaint that Aristarchus was demoting the gods was a serious charge, and got Socrates a cup of Hemlock.

Ptolemy and Aristotle chose the more politic view point that Earth and its rulers were the center of the universe and this view held sway until Copernicus revived the Heliocentric view of Aristarchus in the 16th century. Copernicus was able to take advantage of the weakening Catholic Church's authority caused by the Reformation Movement in Germany, where his book was printed. He also took advantage of the intellectual aftershocks of the discovery of the new world by Columbus, who had used the marvelous new Chinese invention, the magnetic compass.

Beyond Einstein: GEM Unification

Johannes Kepler, the solver of the orbits.

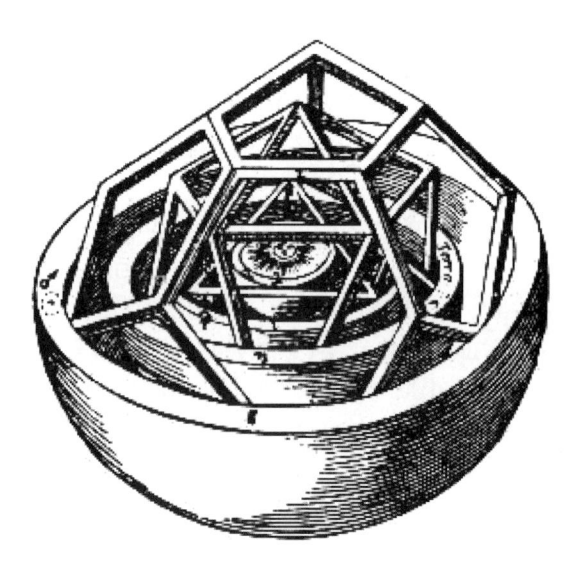

Kepler's original concept of the solar system as nested Platonic solids.

Kepler seized upon the work of Copernicus, and not waiting for absolute proof, solved for the orbit of Mars and the other planets around the Sun. He was motivated by the belief that the Sun was associated with God, about whom all things moved. He was also transfixed by the idea that the orbits of planets were somehow a nested set of Platonic solids. The orbits were nearly circular, but were actually ellipses. But, what force, Kepler wondered, caused the planets to move around the Sun, and the Moon around the Earth? Inspired by the success of the compass, and the analysis of electricity and magnetism by Gilbert, who proposed that the whole Earth was a great magnet, Kepler proposed that magnetism moved the planets around the Sun.

Dr. William Gilbert and his big magnet.

Magnetism was a well known force that acted across empty space. Kepler also found that the periods of the planets were proportional to the distance away from the Sun. The farther from the Sun, the slower the planets moved. But what were the laws that made them move exactly? On Earth the motion of everything was complicated by friction, but in space this friction was absent. Kepler also wrote a book about optics, identifying the decrease of a candle's light with distance as following the law of the inverse square of the distance.

Gallieo, a contemporary of Kepler, studied not only the heavens with his new telescope but also the movement of bodies nearer Earth. He discovered that uniform straight line motion, as on a sailing ship in smooth seas, could not be sensed or detected. He also discovered that the weight of objects was proportional to their mass and worked out the laws of motion for falling bodies. Being a college professor at Pisa, it is said he took advantage of his location to roll two different sized cannonballs off the leaning tower of Pisa to prove that they fell at the same rate. The fact that all bodies, disregarding air resistance, fell at the same rate was an astonishing but useful piece of knowledge. He took this knowledge further and worked out the mathematics of the parabolic curve to chart the course of flying cannonballs, so their ranges could be predicted accurately. He also pointed his new telescope to Jupiter and discovered a solar system of Copernicus in miniature. He saw that the Milky Way was a vast cloud of stars in uncountable millions. He also noted, perhaps more importantly, that the planet Venus exhibited phases like the Moon, showing that it orbited the Sun.

All science is political, and regardless of its truth, is either embraced or resisted in its time depending on the mood of the-powers-that-be. Tycho Brahe, an astronomer and court astrologer as well, promptly adapted his cosmology to make Venus orbit the Sun but   of the planets orbit Earth. His answer to Galileo's observations was the same as the ancient

Beyond Einstein: GEM Unification

court astrologer/astronomer in the presence of Pharaoh, who, when asked what was the center of the universe, answered immediately, "it is of course, you, great Pharaoh." Galileo experienced furious opposition to his interpretations of what he saw through his telescope. His views sounded to the Church like the teachings of Giordano Bruno, who had taken the works of Copernicus and moved beyond them.

Bruno, a Gnostic philosopher, had gone around Europe teaching that not only did Earth and other planets move around the Sun, but the stars were actually other suns with planets moving around them. These planets, Bruno taught, had people on them also. For these beliefs, and his refusal to recant them, Bruno was burned at the stake in 1500. Galileo, being a wise man, as well as intelligent, recanted his statements and was spared the fate of Brunobut was confined under house arrest until the day of his death. In the same year Galileo died Newton was born.

Isaac Newton was recognized as a genius from an early age. He was the son of a merchant but was soon sent off to become a scholar. England was an island of enlightened peace and prosperity in the Europe of those days. He was intensely religious and wrote far more in his life about the Bible than he did about science. After Henry VIII had separated the Church of England from the rule of Rome, and his daughter Elizabeth had skillfully turned her nation into a major sea power, and presided over the age of Shakespeare, the land was a fruitful garden of intellectual accomplishment and inquiry. In 1657 Newton broke up light in prism into the rainbow and invented the reflecting telescope using the parabola of Galileo to focus the light. Newton proposed that light was made of atoms like matter was and that these atoms of light flew across space. He proposed that the prism bent the light because of its weight, or "gravity" as he called it.

Beyond Einstein: GEM Unification

Isaac Newton, as a young man.

Across the English Channel, in Holland, Christian Huygens had become a "rock star" of science in his own country. He had begun by inventing highly accurate clocks, then cast his keen gaze across the whole field of science, or natural philosophy as it was then called. Huygens proposed the concept of light as a wave rather than as a particle. This involved Huygens and Newton in a legendary debate over the nature of light that continued for generations after them.

Beyond Einstein: GEM Unification

Christian Huygens.

To ocean faring peoples such as the Dutch and British, waves of water and even waves of grain in field moved by the ocean breeze were as familiar as sand dunes would be to those living in desert lands. The motion of the wave, especially the wave that was merely a ripple across the water could be watched and studied in a wash basin. Waves that were small in height passed through one another without disturbance. Waves could bend around obstacles, undergoing a distortion called diffraction. Waves in water changed their speed depending on their wavelength, the distance between their peaks, and the depth of the water. Create a sloping beach in your wash basin, and the shorter wavelength ripples would pass through undisturbed in direction or speed, longer wavelength waves slowing and bending their direction. This to Huygens seemed entirely analogous to the actions of light.

A Danish astronomer named Ole Romer, in 1676, armed with a good telescope at the Royal observatory in Paris, and also apparently insomnia, undertook an obsessive campaign

Beyond Einstein: GEM Unification

to measure the precise times of the occultations of the moons of Jupiter. He discovered that the periods between the occultations varied systematically depending on how far the Jovian system was from Earth, as the two planets moved around the Sun in widely different orbits. After consulting with Huygens, Romer announced a spectacular result, that light itself had a finite speed. Their estimate of its speed was enormous, faster than anything imagined before. The speed was so great, that a beam of light from Earth reached the Moon in one second. It was only across the expanse of the solar system that light's progress could be measured by the clocks of the day. As it would turn out, the speed of light was one of the great constants of the universe, a number that contains in it the workings of the vacuum itself.

Ole Romer.

Newton, however, found himself sitting in an orchard, and watched an apple falling from a tree. This commonplace event intrigued him. He launched into a deep inquiry as to why, as Aristotle had said, that everything sought its place in the universe, and the related question of why the planets seemed to find their places continually circling the Sun. It was a well known concept in that time that the Sun exerted an attractive force on the planets, to make them move in circles. Like the magnetic attraction suggested by Kepler of a magnet to iron, this force diminished quickly with distance.

Beyond Einstein: GEM Unification

Isaac Newton, later in life.

Newton, following the suggestion of others such as Christopher Wren, the designer of the Cathedral of Peter and Paul in London, used an inverse square law, used by Kepler to measure light. Newton also invented the calculus to study the dynamics of the planets as they moved close to and farther from the Sun on ellipses. The whole grand synthesis, of the laws of motion, of the law of gravitation and beautiful mathematics of calculus, that Newton invented to express it, was contained in his Magnum Opus: The *Principia Mathematica* of 1687. This book massively altered the way all science and human thought was done.

Beyond Einstein: GEM Unification

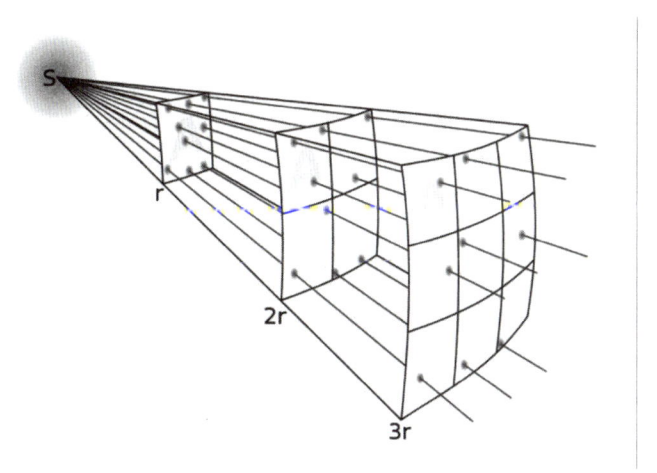

The inverse square law of light.

The *Principia* was key to solving many other problems in physics but most importantly unified the dynamics of the heavens and Earth. The apple falling from the tree and the planet Jupiter orbiting the Sun both obeyed the same laws of gravitation and motion.

Newton found that every particle of matter in the universe attracted every other particle with a force that varied as the inverse square of the distance between them and also depended on the product of their masses. Newton also showed mathematically that spherically symmetric bodies of matter, such as stars or planets, attracted similar spheres as if all their mass was concentrated at the center of each sphere. This meant that collapsing shells of matter could form a star or planet while the center of their mass followed the same orbit. He also explained the tides, which are caused by the variation of gravity from other bodies across a planet's surface.

Because gravity varies as the inverse square of the distance, the gravity of the Sun and Moon is stronger on the side of Earth facing them than it is on the side facing away,

Beyond Einstein: GEM Unification

so that the oceans are pulled towards the Sun and Moon and form tides. This tidal force tends to squeeze Earth on the sides at right angles to the Sun and Moon and tries to elongate Earth along an axis through its center to the center of the Sun or Moon.

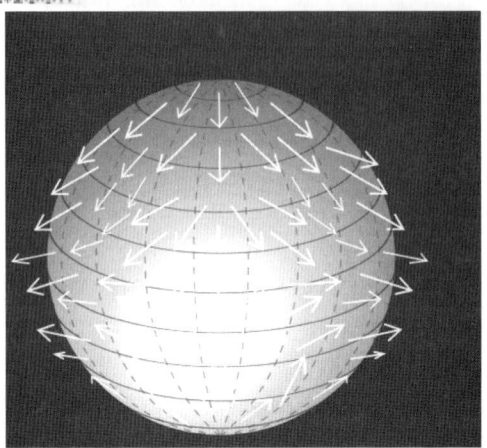

Tidal forces on a planet in orbit around a star.

Newton also made inquiry into cosmology. He quickly found that the cosmos might be unstable to the point of collapse due to his new invention of gravity, because everything attracted everything else. However, he recognized this might get him into trouble with the Church of England, so he did not publish his findings. Newton was religious but unorthodox in his beliefs, being a monotheist rather than a Trinitarian. Therefore, to avoid any inquiry into details of his beliefs, he dodged this confrontation, having learned from the examples of Bruno and Galileo. So Newton left cosmology to Einstein.

Part of Newton's legacy was that he not only solved great problems but developed a conceptual framework to define problems so they could be solved. His laws of motion, for instance, are in actuality a method of bookkeeping. The laws are familiar:

Beyond Einstein: GEM Unification

Bodies at rest or in straight line motion tend to remain in this condition. This is called the law of inertia.

For every action there is an equal and opposite reaction. This law conserves momentum.

Every body attracts every other body with a force proportional to the product of their masses, that is, one mass times the other, and varying inversely by the square of the distance between them. This is Newton's law of gravitation.

One important detail must be added to the last law, the law of universal gravitation. This detail is that a universal constant, G, must be found to multiply this product of masses divided by the inverse square distance. "Big G" is one of the most important known constants in the universe.

The gravitation constant G condenses into one number a complex set of unseen actions in the void that must occur for two particles to attract one another from a distance. As we will see later, in G the heart of unifying EM and gravity lies.

Newton was criticized in his time for introducing the concept of action at a distance, the idea that bodies could act on each other without touching. It seemed a spooky occult-like idea. However, Newton knew that static electricity and magnets could demonstrate this action at a distance on a tabletop. So it can be said that electricity and magnetism, by being so demonstrable, were the midwives of the law of gravitation.

It was known that the inverse square law applied to many other things, such as light from a candle, and later, the actions of poles of a magnet on each other as well as on the charged bodies.

Beyond Einstein: GEM Unification

The powers-that-be were pleased. The highly successful and powerful King of France declared himself the "Sun King," about whom everything orbited. A sense of triumph of the human intellect over the cosmos swept Europe. "Newton has explained the universe," went the cry. A deep feeling that everything could be explained rationally and its course determined became prevalent.

Louis the 14th King of France: "The Sun King."

By defining the problems of moving bodies in the void, and then inventing calculus to solve these motions, the means to solve almost every conceivable problem of the day was laid before them. Newton made England the center of science from that day forward.

Chapter 4. Magnus and Electra

"I am busy just now again on electro-magnetism, and think I have got hold of a good thing, but can't say. It may be a weed instead of a fish that, after all my labor, I may at last pull up."
Michael Faraday

"At quite uncertain times and places, The atoms left their heavenly path, And by fortuitous embraces, Engendered all that being hath."
James Clerk Maxwell

Electricity and magnetism have been known since ancient times. The Egyptians were blessed with an electric eel that inhabited the Nile called "thunderer," apparently connecting it to thunder and lightning. The marvelous eel, instead of having muscles created merely for swimming, was cleverly endowed with modified muscle organs that generated electricity. It is quite capable of generating a strong shock to anyone foolish enough to grab it. In a Nile full of predators, this talent gave much survival value. A similarly talented fish swims in the Mediterranean, and gave us the name "torpedo" from the Greek word "torpor" or numbness, which

Beyond Einstein: GEM Unification

it induced in the hands of the unlearned fisherman plucking this particular fish from his net. The Romans also noted that poking such a fish with bronze shafted trident produced the same result as grabbing it with your hand.

Electric eel, no joke.

Thales of ancient Greece noted that lodestones, natural magnets, obtained from the province of Magnesia, attracted iron. The substance amber, when rubbed against dry fur was noted by Thales to produce a similar effect with feathers and other light materials. He considered that the amber became magnetic upon rubbing, hence the two forces, magnetic and electric, were described from the ancient times as being unified. The Chinese scholar Shen Kuo described magnetism and electricity in the book *The Dream Pool Essays* and even before this magnetism is described in the book, *Book of the Devil Valley Master*. An Indian physician also described the use of the lodestone in surgery, probably to help extract iron arrow heads from the wounded. The Chinese invented the magnetic compass in the 1200s and its use spread to Europe, where it made possible the great voyages of Columbus and Magellan.

Beyond Einstein: GEM Unification

Despite these documents electricity and magnetism remained of little intellectual interest until the English scientist William Gilbert studied them and published his analysis in the book *De Magnate* in 1600. He found that magnetic and electric forces were quite distinct. He analyzed the action of the compass and identified the fact that Earth itself was a great magnet. Magnets attract iron continually, and thus find a place holding up report cards and bills on the refrigerator. Electric force, generated by static electricity, electric charges at rest, attracts pieces of paper and fades quickly. Electricity is also dynamic, creating a strong shock and spark of light, when one walks across the carpet on a dry winter evening, to lock the front door of one's house. Gilbert, from a time when scientists were known as natural philosophers and thus closer to poets, gave us the names magnetism and electricity, from their ancient origins.

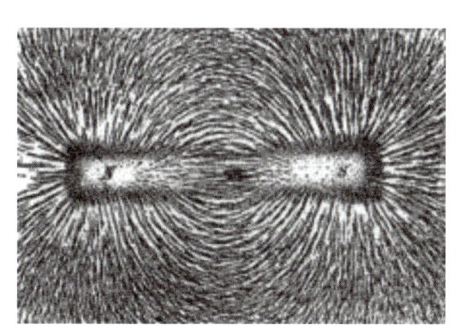

The lines of force around a magnet seen in iron filings.

It was quickly found that, like light and gravity, electric and magnetic forces follow an inverse square law of attraction and repulsion.

As electric and magnetic forces were studied it became apparent, that while distinct, they had close mathematical similarities. Both magnets and electricity-charged objects attracted things with forces that diminished strongly with distance. Magnets and electrically charged objects both

Beyond Einstein: GEM Unification

displayed a complex dance where like poles and like charges repelled each other and unlike attracted. Thus, it was also found that bar magnets, when split, would form another set of paired poles, while isolating negative and positive electric charge was easily done. It was understood, with awe, that nature generated a vast magnetic field in the core of Earth itself, and thus a compass needle could be used to guide a ship during the day when the stars could not be seen.

Ben Franklin, one of the first American scientists, discovered that nature generated vast amounts of static and dynamic electricity during thunderstorms. His discovery, along with its immediate practical invention of the lighting rod, made him famous in Europe and made the success of the American Revolution possible. This occurred because Franklin could be received at the French court as a famous figure, thus securing the French as allies against the British.

Beyond Einstein: GEM Unification

Ben Franklin.

Michael Faraday was a commoner and a member of an obscure Protestant sect known as the Sandlins. He rose from being an assistant to the English scientist Joseph Dalton to being a well known scientist in his own right. He found he could create magnetism by electricity and electricity by magnetism, achieving rudimentary unification. He built the first electric generator, and the first electric motor, converting mechanical motion into electricity and then back again. He invented the concept of the force field and lines of force, representing the vector, from the Greek word for "arrow," to represent the value of the electric or magnetic field of force. He also made the remarkable discovery that light had a polarization direction, and that this polarization could be made to change by magnetism when light moved through certain materials. This led him to consider that light itself was a manifestation of electric and magnetic fields.

Beyond Einstein: GEM Unification

Michael Faraday.

Maxwell, a very religious Scot, journeyed south to teach in London and enjoyed the company of Faraday, though 40 years his junior. Maxwell was fascinated by light, and had taken the first color photograph by imaging a brightly colored tartan ribbon through filters of three different colors. When he projected the images of the ribbon taken through the three colored filters onto a wall, and focused them, he demonstrated that the original color image of the ribbon reappeared to the human eye. Color imaging was thus born. He contracted smallpox and was at death's door but somehow, in spite of this experience or perhaps because of it, completed the four great equations of electricity and magnetism that we now know as Maxwell's equations.

Beyond Einstein: GEM Unification

James Clerk Maxwell as a young man.

He completed the unification of electricity and magnetism just as the United States had disunited and was plunging into the Civil War. Maxwell's equations can be understood with the help of two concepts concerning force fields and lines of force. One is the concept of radiance, the idea of a force field where all the lines of force radiate from a central region outwards in every direction, like light radiates from a candle. This is called, by scientists, divergence, but radiance is more descriptive for the nonscientist. The radiance of a force field can be measured by placing a spherical shell around the center of the radiant fields and counting all the lines of force that pass outwards through the shell's surface. We will call total outward flux of lines of force the total radiant flux.

Radiance carries in itself the concept of the sphere and its area. The total energy of the light of a candle, if measured by a sphere coated on its inside with solar cells, is the same no matter how large the radius of the solar powered sphere. This is because the flux of energy of the light of the candle diminishes with distance by the square of the distance. The surface area of the sphere covered on the inside with solar cells, on the other hand, becomes larger as the square of its radius. So the diminishment of the flux from the candle and increase of the area of the solar cell sphere compensate for each other and as a result the net power of the light radiated by the candle is the same always independent of the radius of the sphere.

Beyond Einstein: GEM Unification

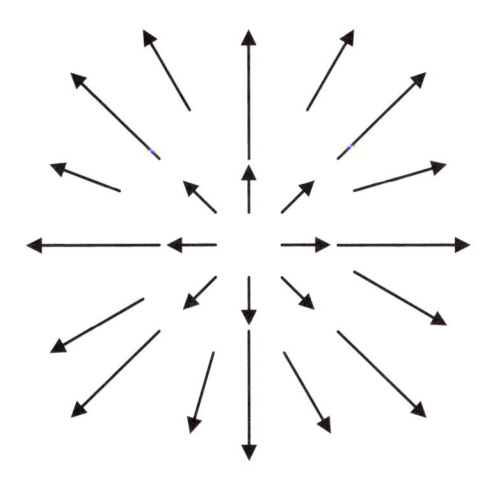

A radiant vector field pattern.

The radiant energy of light is bound up in a vector field called the Poynting vector. The Poynting vector was discovered by John Henry Poynting, the son of a Unitarian minister and a student of Maxwell. The Poynting vector points in the direction of the energy flow in an electromagnetic field and is formed by the crossing of electric and magnetic lines of force.

Another property of a force field is the vorticity or "swirl" of the lines of force. If a fluid flow field has swirl and a small paddle wheel is placed there it will turn. This mathematical quantity is called curl by scientists, but again swirl conveys its more dynamic aspect. Together the swirl and radiance of a force field in a region describe it completely mathematically.

Beyond Einstein: GEM Unification

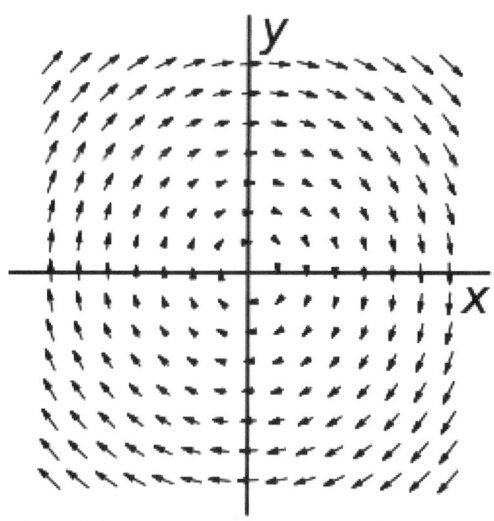

A swirl or vortex pattern of vectors.

Armed with these ideas we can describe Maxwell's equation in words:

The first law is simple, it says the radiance of the electric field in a region is due to the electric charge in the region. By convention, positive charge in a region gives outward radiance and negative charge gives negative radiance. Unfortunately, this convention also means that flows of electric charge is the reverse of the real motion of the electrons, which are the main carriers of electric current, because they are light and move easily compared to the protons that carry positive charge.

The second law was discovered by Michael Faraday and bears his name. Faraday's law says that the change of magnetic flux through a region creates a swirl of electric field around the region.

The third Maxwell equation says that the radiance of magnetic lines of force from a region is always zero. This means there is no isolated magnetic charge like there is an

Beyond Einstein: GEM Unification

electric charge. As was quickly discovered when magnets were made from lodestones, a magnet always had two magnetic poles. One could make the magnetic long and skinny and thus have one end act like an isolated magnetic pole, but the other pole is always connected to it. We will find that this means the magnetic lines of force are actually endless vortex lines, whereas electric lines of force are simpler arrows going from one place to another.

The fourth Maxwell equation is what made Maxwell famous. To see this we will look at the equation before Maxwell fixed it. Before Maxwell's correction the equation says simply that the flux of electric current through a region creates a swirl of magnetic lines of force around the region. This means that magnetic lines of force circle a wire carrying current. This was called Ampere's law. However, Maxwell recognized that Ampere's law was incomplete, or, rather, that it needed to be corrected. A problem arises when one considers the radiance of a field that is all swirl. The swirl is vorticity and the lines of vorticity are endless loops. Therefore, a field that is all swirl can have no radiance, for the lines of force of a radiance field must end inside a sphere.

If one combines the radiance of both fields, the swirl of the magnetic lines, and the flow of current, the radiance of the swirl must equal mathematical zero. Vortex lines are endless loops; they can have no radiance. A swirl defines a vortex line. The swirl of the magnetic lines of force cannot have radiance. Therefore, the swirl of Ampere's law always equals zero. Here, Maxwell saw an error. He saw that Ampere's law was incomplete as an equation.

However, the radiance of an electric current in a region, the other half of Ampere's law, does not always equal zero. If one created a ball of charges of all the same sign, they would repel each other and explode. The like charges would

Beyond Einstein: GEM Unification

fly outwards in every direction along the lines of electric force that are radiant from the region of charge. Thus, one side of the equation can have no radiance but the other side can. There was the missing piece of the puzzle.

Maxwell corrected the other side of Ampere's Law with a new term, so the radiance of both sides would mathematically vanish. He made sure that the equation conserved electric charge with the flow of current. He added the new term to the source of the swirl of magnetic lines of force. This new term was the rate of change of electric flux through the region. Now the change of magnetism could make an electric field, and the rate of change in the electric field could make magnetism. The two fields could feed on each other, like two serpents winding around each other.

It seemed like a small thing at the time, like adding a period to a sentence. However, as Maxwell studied this new term that balanced the equations, he realized he had forever changed the world. By the time the American Civil War was drawing to a close, two great associations had been unified, one was the United States, and the other was three things: electric fields, magnetic fields, and light.

The final equation that completed the set of four equations, called henceforth Maxwell's equations, showed that the swirl of the magnetic lines of force around a region was due to the flux of electric current through the region plus the rate of change of electric flux through the region. This meant that the rate of change of electric flux could give rise to swirling magnetism and the swirling magnetism, as it changed, could give rise in the second of Maxwell's equations to a changing swirl of electric lines of force. The void itself was now a machine full of gears and levers, it seemed, or, as many imagined, a fluid, called after Aristotle, the aether. The chain of interlocking swirls of electric and magnetic fields acting in harmonic rates of change with time

Beyond Einstein: GEM Unification

could move as an independent entity, free of wires or amber or lodestones; it could dash across the cosmos at the speed of light. Maxwell had discovered the harmonic electromagnetic wave. Maxwell had shown forth the mechanism of sight itself, light.

Maxwell had unified the two fields of force known since the ancients, electric and magnetic fields, but Maxwell's work showed that they could combine to form a third field.

The third field is called the Poynting field, after John Henry Poynting, and it is this field that carries energy and momentum in a harmonic wave across the void with a beam of light. Poynting was the son of a Unitarian minister and a student of Maxwell's. The poets are pleased to note that the Poynting field, composed of electric and magnetic field lines that are crossed to each other, points always in the direction of the light beam's travel.

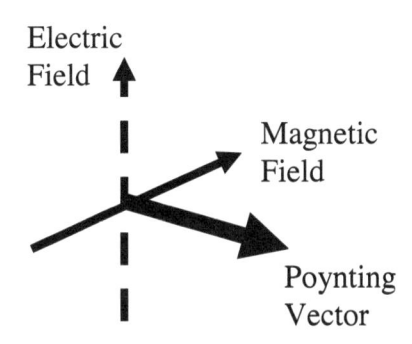

The Poynting vector of electromagnetism.

Beyond Einstein: GEM Unification

John Henry Poynting.

All electric and magnetic fields exert forces. The electric field pulls or pushes charges along its lines of force. The magnetic field, being a vortex field, bends the paths of moving charges continuously to run in circles around its lines of force. This force was discovered by Lorentz of Holland and is now called the Lorentz force. It shows that magnetic lines of force are actually lines of vorticity, and that electrons in the vacuum will move according to that vorticity. So the electric and magnetic fields each create forces on charges. The third force field, electromagnetism, is intermediate between gravity and electromagnetism.

Beyond Einstein: GEM Unification

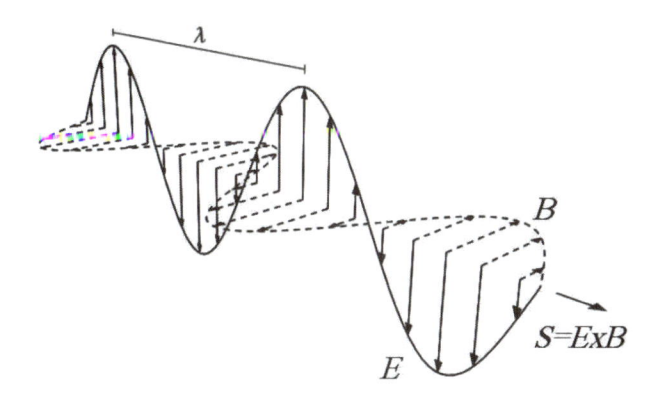

The Electric, Magnetic (B) and Poynting Fields of an EM wave.

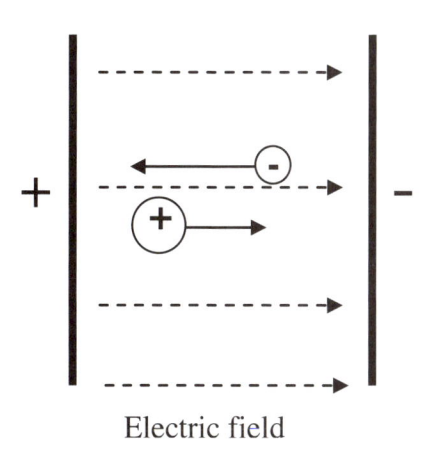

Electric field

The reaction of an electron (-) and proton (+) to an applied electric field; they are pulled in opposite directions.

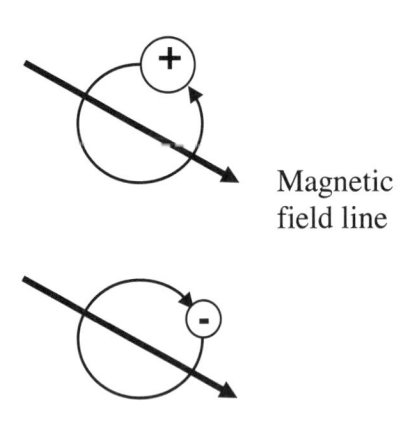

Beyond Einstein: GEM Unification

The paths of an electron and proton in a magnetic field, in which they orbit the field lines in opposite directions.

The Poynting field creates the third force of electromagnetism: radiation pressure, the pressure exerted by light. The Poynting force can be understood as using both electric and magnetic forces simultaneously, for it points in the direction of the light beam's path, and when it encounters an electron or proton, it affects them as a combination of electric and magnetic fields. The electric field of the beam of light makes a charge move up and down in its harmonic field across the path of light. The magnetic field bends this motion due to the electric field of the light beam so that the up and down motion is converted to a push in the direction of the Poynting vector. This combination of electric and magnetic forces and motions is the source of radiation pressure, the pressure of light. For this reason, a great aluminized plastic sail can be unfurled in space to catch the rays of the Sun and can drive a spacecraft to Mars. On the atomic scale a ray of light can strike an electron in a crystal of silicon and push it across a junction to another part

Beyond Einstein: GEM Unification

of the crystal, thus converting the energy of the light into electrical energy. This photoelectric effect was first analyzed correctly by Einstein, who saw that the Poynting field of the light affected each electron like a particle. For this discovery Einstein won the Nobel Prize.

The Poynting Force at work: The Pillars of Creation are carved from cosmic dust by radiation pressure.

Beyond Einstein: GEM Unification

We can see now that the electromagnetic force is actually three distinct forces. There is the attraction and repulsion of two electrically charged balls of cork. There is the attraction or repulsion of the poles of two magnets. Then there is the Poynting force, the pressure of light on matter, so that in the vacuum space, one can raise golden sails and cruise among the planets without rocket fuel. By the Poynting Force the light of stars carves the galactic dust into the pillars of creation. All these forces can be seen in complete isolation from each other, yet all have their source in the unified theory of electromagnetism. All follow the same mathematical law, that of inverse square of the distance between them, but each is independent of the other.

Maxwell had unified the electric and magnetic forces and their dynamics. Henceforth, they were referred to no longer as electricity and magnetism, but together as one word, electromagnetism. Not since Isaac Newton had such a feat been done. Sadly, Maxwell's brilliance was cut short by abdominal cancer a short ten years after his triumph.

He made many other great contributions to science in his short time on Earth, most principally in the new science of thermodynamics. The concept of matter being made up of atoms and molecules had become a science, and with it the study of entropy or disorder.

With the discovery of the atomic nature of matter came the discovery of the atomic nature of electricity. With the discovery of the electron as the fundamental carrier of charge a new mystery presented itself. Like charges repel. One could not shrink a copper sphere with many electrons on it without its feeling stress and then exploding because of the repulsion of the electrons from each other. Sooner or later, at some small radius, the strength of the copper would give way. However, if the electron was itself made of charge, what held it together? People decided not to pursue this question, considering that the electron was somehow indivisible, an atom of electricity, and therefore could not explode. However, if atoms were made of plus and minus charges, they would be held together by electromagnetism. Thus, the atomic nature of matter made great sense in the context of EM.

Maxwell worked out the spectrum of energies of atoms of molecules in a gas at a given temperature. The only way to analyze such a number of molecules in any given everyday volume was to use the laws of statistics. His answer was an exponential function in the energy divided by the temperature of the gas. This meant that molecules or atoms with energies above a characteristic energy associated with the temperature of the gas would become increasingly scarce. This law basically defined temperature as an average energy of the molecules in a gas. Written in terms of gas molecule velocity the statistics follow the familiar "Normal Curve" seen in many other areas of statistics.

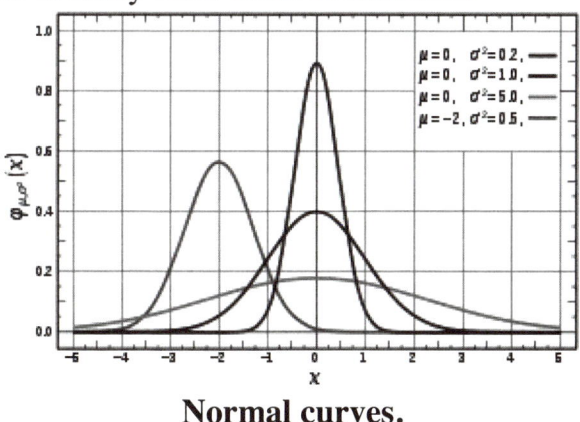

Normal curves.

Basically, for a gas, the statistical spread of atoms over energies depended on the temperature, an average energy of the atoms or molecules. At a very low temperature called absolute zero, all the molecules would stop moving and rest at zero kinetic energy. As the temperature of the gas rose, some of the atoms would pick up speed and form a statistical spread of energies, a spectrum, that would be nearly constant until the average energy was reached, but then drop off rapidly in number. Finally, at infinite energy, no molecules would exist at all. Entropy, the spread of the spectrum from zero to some high energy, thus increases proportionally to temperature.

Maxwell later in life.

The discovery of this thermal spectrum by Maxwell greatly increased the understanding of thermodynamics and, together with his unification of the magnetic and electric force fields, paved the way for the next great revolution in physics.

Maxwell's advances of EM theory led to the invention of radio and all other forms of wireless communications technologies, but also to tremendous scientific advances. This followed because harmonic EM waves crossed the universe and reached into the atoms it contained.

Matter is fundamentally electromagnetic in nature. In the simplest example of hydrogen the electron orbits the proton like a planet orbits the Sun because they are of opposite

Beyond Einstein: GEM Unification

charge. When the hydrogen becomes hot the electrons vibrate as they orbit and act like little electromagnetic antennas and emit electromagnetic waves that we see as light. When they are cold they can absorb light in the same wavelengths. In more complicated atoms that make up everyday existence, electrons move in large numbers around the positively charged nucleus like a cloud and the outer electrons sometimes wander off from their atoms. Electricity, both static and dynamic, arises from the fact that the electrons in atoms are not always held closely, but sometimes the rubbing of amber against animal fur can cause electrons to stick to one or the other and create a difference of charge. In thunderclouds the change of state of water from vapor to liquid and solid apparently separates electrons from their atoms and causes the cloud to turn the vast releases of thermodynamic energy into vast amounts of electrical energy. The lightning bolts from this vast electrical generator strike the ground, and if the soil is full of iron ore as in Magnesia, they make lodestones. When the storm is over, the water droplets in the clouds catch the rays of the Sun and break it into its different harmonics, perceived by the eyes as colors, so that a rainbow appears, just as Newton predicted.

Three men's portraits hung in Einstein's study: Newton, Faraday, and Maxwell. These great unifiers urged Einstein on to further unify the great forces of nature.

Beyond Einstein: GEM Unification

Chapter 5. Atoms of Light

"Science cannot solve the ultimate mystery of nature. And that is because, in the last analysis, we ourselves are a part of the mystery that we are trying to solve."

Max Planck

Max Planck, a conservative German professor, set off a revolution in 1900, at the dawn of a new century. He had been pondering the problem of the EM fields in a box. His was a simple problem if one only looked at the wave fields of Maxwell and one considered the inside of the box to be made of perfect mirrors. Like a plucked harp string, the waves in the box would be harmonics of the lowest vibration that could fit in the box. Because of the simple additive nature of EM waves, the waves could be harmonics added one on top of the other like pages of a book. Onward the waves would go, upwards in frequency, to infinity, like a stairway to heaven. If one drilled a hole in the box and looked, one would see only darkness because the mirrors were perfect, neither absorbing nor giving energy to the waves. So this was marvelous, orderly, but sterile. The energy of the box would be as it started, zero. But this was not real. Suppose one made the box of common iron and then heated it so the iron turned red hot?

This experiment was easy to do and the results were remarkable. Instead of revealing darkness, the hole in the box glowed brightly, more brightly than the surface of the hot iron, and revealed a dense pattern of EM waves inside. One can do the same thing by cupping one's hands together and feeling the heat of the EM field that forms inside. You can create your very own EM field, and this is a good biofeedback method if one wants to increase the blood flow to your hands. Just concentrate on increasing that EM field.

Max Planck.

When the iron box was heated the light coming from the hole in it became brighter, and to human eyes turned different colors. The color began as red, then turned orange as the box was heated further, then yellow, and then white like the Sun just before the box melted. Scientists all knew by then, because of Newton, that white light was a mixture of colors. This, according to Maxwell, was a field of EM waves of different frequencies. But what determined the statistical spread of light energy over the different colors?

Measurement of light energy had advanced in the 18th century, allowing a statistical distribution of energy over the spectrum wavelength, or color, to be measured. The spectrum of the light from the white hot iron box looked similar to the distribution of energy predicted by Maxwell for atoms in a hot gas, but was different in many details. It was as if the light waves in the box were behaving like a gas of atoms composed of pure light.

According to Maxwell's equations the box should fill with different frequencies reaching up to infinity. Theory predicted that the high frequencies would dominate and finally at some critical temperature all the light would become ultraviolet, leading to an "ultraviolet catastrophe." It is a tribute to the seriousness with which human beings regard abstractions that this predicted catastrophe was considered possible despite the fact that people had been heating iron boxes for centuries and no catastrophe had appeared.

Max Planck attacked this problem by first finding a mathematical function that fit the energy spectrum of light. He found a simple function that did this and then set about to find the theory that produced it. What he found astonished him and harkened back to the old debate of Huygens and Newton over the nature of light.

Beyond Einstein: GEM Unification

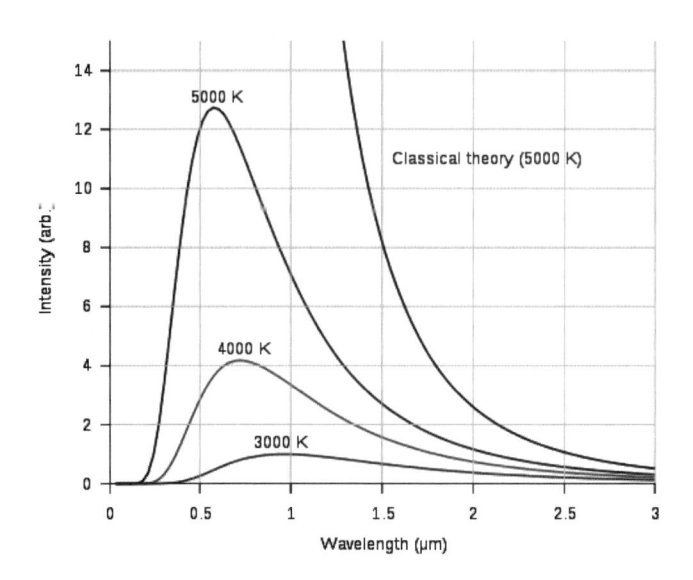

The Planckian spectrum with classical theory shown for comparison.

Max Planck found the only way to reproduce the spectrum of light from the heated box was to treat the light not as waves but as atoms or "quanta" of light energy. The EM field, supposedly a determined solution of Maxwell's equations, was displaying entropy like matter. It was a collection of things, not a single individual entity. The source of this entropy or randomness was the atoms of the walls of the iron box. The atoms of the walls could be viewed as a collection of vibrating springs. They vibrated randomly, excited by the heat of the box, and they were passing this randomness smoothly into the EM field because it was itself a collection of atoms, atoms of light. To pass energy back and forth with atoms, light must also have atomic properties. Newton, centuries before, had been right.

Beyond Einstein: GEM Unification

Huygens had been right, too. Light was obviously a wave, with frequency and wavelength; it did everything a wave field should do. Planck devised a concept of the photon, a quantum or packet of electromagnetic radiation. The higher its frequency, the higher the energy the photon carried. This was obvious, in that infrared light is the carrier of radiant heat away from hot objects, but it does not cause sunburn. Infrared light, being of lower frequency than visible light, carries less energy per photon. Ultraviolet light, however, of higher frequency and shorter wavelength than infrared, burns the skin because each photon has enough energy to disrupt molecular bonds in the proteins in the skin. The constant that determined the amount of energy per frequency in the photons was identified by Planck as a new fundamental constant of nature. The constant had units of vibration or spin, momentum bouncing back in a box or moving within a circle of a given radius. This quantity is called generally "action" and it had been seen before in physics.

Newton's laws of motion can be formulated from the economies of. The product of the energy of a system and the time it experiences dynamics is called action. It is an odd quantity, with the same units as spin. When action is minimized, or, when the system evolves in spacetime along the shortest path, then the world moves according to Newton's laws. Minimizing action is a compact way of writing down all of Newton's laws in one line. The mathematics of how nature finds this shortest route through spacetime was originally worked out by Newton and is one of the most powerful mathematical tools of physics. The action that is minimized was formulated by the Frenchman Lagrange, and is a peculiar quantity of the kinetic energy of an object, its energy of motion minus its potential energy. Lagrange's formulation says that in a collection of systems, nature seeks a balance between potential and kinetic energy as all the particles move. If we return to our pearl in a bowl and put the pearl at the upper lip of the bowl, it has maximum potential energy. When we release it so it rolls down the side of the bowl picking up speed, it changes potential energy into kinetic energy. When the pearl reaches the bottom of the bowl, it has maximum speed and thus maximum kinetic energy and at the same time its potential energy is minimal. It then rolls back up to the lip of the bowl on the other side, neglecting friction, stops, having traded in its kinetic energy for potential energy and then the cycle begins again. Following Newton's laws, the pearl has conserved the total energy at all times, but it has done something more profound, it has minimized its action over time.

Minimum action says that light rays travel in straight lines in spacetime and that all objects maintain their state of motion; it says that action and reaction are paired. The abstract idea of minimum action runs the world we see in front of us. It says the cosmos, in its dynamics, does so on a budget. Even miracles are metered out carefully. Thus action was traded in nature like pennies, like matter itself, in identical units.

So Planck had found a unit of action that all light obeyed, that underlay reality. This constant bears his name, "Planck's constant," and gives a glimpse into the unseen principle that determines all that is seen. But if light carried energy in units, how did it interact with the miniature solar systems of the atoms, which moved continually with electrons circling their nucleus?

Einstein made an important correction to Planck's theory in 1913. Planck had thought the vacuum would lie still at absolute zero temperature. Einstein pointed out that the ZPF was non-zero. The vacuum had to vibrate continually. This would have enormous implications later. Einstein, in effect, has laid the groundwork for field unification even as he charged off in another direction.

Niels Bohr studied the simplest atom that embodied the cosmos from the moment it was created and made up three quarters of its present substance. This atom was hydrogen. Hydrogen, "water former," is just one proton and one electron. No simpler system could be found and yet it was the brick with which the complex cosmos had been constructed. Hydrogen, when excited by heat or electricity, gives off light, but only in specific wavelengths. Bohr realized that this meant that energy was quantized even in the hydrogen atom, in order to emit and absorb energy in the form of light quanta. The electron orbiting the proton could not move continually from one orbit to another, but had to jump quantum steps. Only in the limit where the electron had so much energy that it was about to leave the proton's electric embrace and roam the cosmos freely could its jumps be so small that they looked like smooth transitions.

Niels Bohr.

Beyond Einstein: GEM Unification

A property of electrons is that, while bound to the proton in hydrogen, or a nucleus in more complex atoms, the electrons could receive so much energy, such as when the gas is at very high temperature, that they would leave and roam freely. When the electrons left they leave behind a positively charged ion, and so the gas is called ionized. The freely moving electrons do not venture far; they are still negatively charged and the ions are still positive, but their relationship is no longer governed by quantum mechanics. When a gas is ionized it glows and conducts electricity just like copper, because the electrons are free to move. This gaseous state is called a plasma, from the Greek word for something that can be molded and shaped, the same root as the word plastic. Plasmas are the stuff of stars and the lightning, the neon sign and the shimmering aurora. They are full of entropy, they are full of light, and they have large numbers of electric charges moving freely. From the study of plasmas we will gain an important clue as to the unification of gravity and EM.

The closely bound states of the electron in hydrogen fascinated Bohr. Electron means brightness in Greek, a poetic name for a particle that is agile and heavily charged. Whenever an electron moves it wrinkles the electric fields around it and must, by Maxwell's equation, create EM radiation. The proton does this also, but since it is much heavier, it is the electron that causes most of the motion and makes most of the light. The orbital motion of the electron around the proton and hydrogen should create light waves, and theory agreed with experiment to that degree. However, the light waves would be so powerful they would carry away all the electrons' energy in a few orbits. The electron should orbit a few times then collapse into the proton, echoing the ultraviolet catastrophe before it, one of these predicted catastrophes of science that contradicted all experiments.

Beyond Einstein: GEM Unification

Bohr postulated that the electron could give up energy only in quanta, in the form of light. This meant the electron's orbital behavior must be quantized also. The simple application of this idea led to a breakthrough. The model of quantized orbits with light emitted only when the electron jumped between them gave the line spectrum of hydrogen almost exactly. This was scientific triumph and a conceptual breakthrough, and meant that just as light was part wave and part particle, so matter had the same particle wave duality. The orbits of electrons could now be explained like being waves on a plucked string, so that only waves that resonated with the orbit size would exist. However, the great caravan of science moves on and soon the sense of triumph faded. A thousand new questions emerged. As Newton said, "the larger the island of knowledge, the longer the shoreline of wonder."

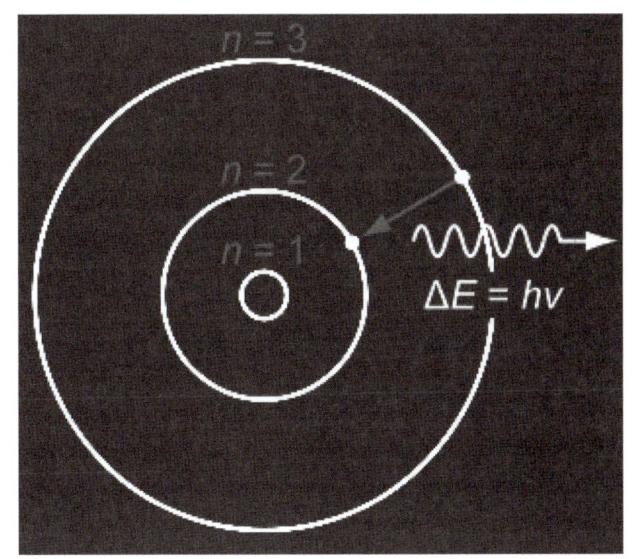

The Bohr model of the atom with quantum levels.

Beyond Einstein: GEM Unification

Now that the position of the line spectra of hydrogen could be predicted, what about the relative brightness of the lines? Why were some lines more intense than others? Why did some lines split into two lines when excited hydrogen was placed in a magnetic field?

To answer these questions, Bohr and his younger protégé Heisenberg abandoned all physical models and created purely mathematical constructs. The Copenhagen school of quantum mechanics, of which Bohr was the leader, dismissed all attempts at visualizing quantum mechanics, claiming it was a fathomless mystery. They even rejected the wave mechanics models of waves fitting on orbits. Instead they put forth a featureless matrix, a tic-tac-toe pattern with numbers instead of x's and o's in the spaces. This matrix was all there was to the model. To Bohr and Heisenberg, *"there was no there, there,"* because in quantum mechanics the human mind had reached the limits of useful thought process. Heisenberg thought only in terms of observables.

Heisenberg put forth his famous Uncertainty Principle that said the closer you tried to look at a particle's path the more vibrations around that path it exhibited. It was the quantum principle that everything vibrated and the vibrations rode on vibrations so that the closer you looked the fuzzier things got.

To understand Heisenberg Uncertainty one can consider a single quantum particle to behave like a gas. If you restrict the gas in a box and squeeze it to localize it, it will get hotter because you have compressed it. That is the price you pay to make the gas occupy less space. The gas molecules will gain energy because they are now confined to a small volume. So it is to try to bring a particle into a finer focus is to do work on it, to compress it, and in response it will gain energy. To even look at an event meant to disturb it, to be part of the event. The watcher and the watched were forever part of the same system.

Werner Heisenberg, the father of "uncertainty."

However, David Hilbert, a brilliant mathematical physicist in Germany and collaborator with Einstein, reportedly commented to Heisenberg that, when he had encountered matrices like those seen in quantum mechanics, a field equation had always generated them. This equation would have wave-field solutions. However, Heisenberg and Bohr scoffed at this suggestion. Clearly Hilbert was an old graybeard addicted to old style thinking where things where supposed to make sense, but this was the new reality of quantum physics.

Beyond Einstein: GEM Unification

David Hilbert, the master of mathematical physics.

Edwin Schrodinger labored under no profound convictions like Bohr and Heisenberg. To Schrodinger the idea that quantum mechanics would be forever a fathomless riddle was, in a word, ridiculous. It was a problem that could be solved. So he wrote down an equation that tied together everything that had been seen. When he sensed hostility from the Copenhagen school about getting it published, he sent it to Einstein, who looked it over and recommended it to a journal. It was published with Einstein's endorsement in January 1926. Einstein was skeptical of quantum mechanics, despite having helped to gain it acceptance. To Einstein this Schrodinger equation looked like a good way to throw a cherry bomb into the courtyard of the Copenhagen School.

The Schrodinger equation, as it came to be called, made quantum mechanics into a field theory like Maxwell's equations. Instead of electric and magnetic fields it was a field called ψ. However, ψ was not a field like temperature or pressure; it was a field of probability. Suddenly quantum mechanics made a strange sort of sense. The matrix of Heisenberg was found to follow directly from the

Schrodinger equation. Bohr and Heisenberg were horrified; Einstein was delighted.

Erwin Schrodinger.

Many events intervened after these years of creative ferment and the sharp thunder of war swept over Europe. The cream of world physics gathered at Los Alamos, New Mexico to create the atomic bomb. Afterwards a young physicist named Richard Feynman unified quantum mechanics with EM. He called his theory QED or Quantum Electro-Dynamics.

Feynman produced a beautiful theory in which the wave fields of the electron and the EM field unified and moved according to the principle of least action. Feynman found a remarkable thing: The EM field producing forces between charges and current could be broken down into a pattern of photons. The forces of the EM field, even pure electric and magnetic forces, could be written down as radiation pressure

Beyond Einstein: GEM Unification

from photons. Gone was action-at-a-distance, gone was the force field, and replacing it was a cloud of radiation photons, hitting electrons and transferring energy in "photo-electric effect" collisions as first proposed by Einstein.

Richard Feynman, the merry prankster.

In an example of inverse square law, when two bright objects are placed in a dark box, they will repel each other with mutual radiation pressure.. Thus was Feynman able to make charges as bright emitters of photons that pushed on other charges. This accounted for the repulsion of like-charges. By adding polarization of the photons, attraction between unlike charges could also be explained. However, in the mystery of quantum mechanics, the photons making up these fields could not be seen directly. The photons were invisible, or rather subliminal, but moved everything.

When Feynman and his colleagues unified quantum physics with relativity and electrodynamics, they hoped to address the riddle of the electron, that most perfect of particles. The riddle had become more profound since Einstein discovered relativity.

If one took the mass of the electron to be pure EM energy, then one obtained, through Einstein's relativity, a radius for it. This occurred because to create an electron from a cloud of charge one had to compress all this charge into a small sphere, and the smaller the sphere, the harder this became, because the charge repelled itself by a one-over-distance-squared force law, and all the charge had to be pushed closer together. The electron wanted to explode, not form a particle. The smaller the electron the greater the compression required, and thus more work, which is energy, did this compression require. So the smaller the electron became the greater the mass of the electron became. One could use the observed charge and mass of the electron to find the radius it must be if it was a hollow sphere of pure charge. This made perfect sense in classical EM, since the individual electron could scatter light like a sphere of silver with this same classical radius. This left aside the problem of what held the electron together.

However, rather than solving the problem of the electron radius, quantum mechanics made it more profound. The problem was that the only quantum model of the electron that made sense was that the electron was a perfect point with a radius of zero. This would mean its mass-energy would be infinite, since the charge would have to be infinitely compressed. A point electron would weigh infinity.

The problem was made slightly better by the quantum mechanical vacuum around the electron. The quantum vacuum was full of virtual particles of plus and minus charges. When the electron was shrunk to its ideal zero radius to form a point, the quantum vacuum would be polarized like wax or some other everyday substance, and positive charges would appear according to Heisenberg uncertainty and would be attracted to form a cloud of virtual positive charge near the electron, while virtual electrons, being of negative charge were pushed away and formed a cloud surrounding the positive charge cloud. This made the electron charge appear weaker from a distance, because the quantum vacuum partially neutralized it. For this reason the electron was believed to have a "bare charge"; that is, much larger than its measured charge which includes the vacuum polarized neutralizing charge. In a sense, while relativity had supposedly banished the idea of an aether from physics, it had sneaked back in the form of the rich substance of the quantum vacuum. This may explain why Einstein found quantum mechanics increasingly annoying.

The idea of a higher bare charge for the electron than what was seen in the laboratory did nothing to help the problem of its being supposedly a point yet having finite mass. The quantum vacuum helped only slightly: Instead of infinity it produced a logarithm of infinity. The logarithm of infinity is still infinity, so after several years of hand wringing and failed attempts to make this new "ultraviolet catastrophe" go away, the electron was simply assigned a much smaller radius than before and the matter was dropped with some embarrassment. The assigning of a new much smaller radius was termed "renormalization" and is essentially the statement that at a small enough radius new physics occurs which negates the effect of smaller sizes. It is literally subtracting an infinity from the first infinity to force the answer for the electron mass to come out right. Renormalization pleased few in physics community, and was termed a *"dippy process"* by Feynman. It was like hanging a picture over a hole in the wall so no one would see it. Dirac commented that he *"had heard of numbers being neglected because they were small, but never when they were infinite."*

Beyond Einstein: GEM Unification

However, the fact that physics must change at some small size to make the electron mass finite has been found to be necessary in analysis of the strong and weak forces also, so it must be physical. Somewhere in the deep subatomic scale, physics changes.

With the problem of the finite mass for a point electron requiring such mathematical gymnastics to solve, the secondary problem of what held the electron together seemed like a nonsense question. Why worry about such things when the electron was part particle and yet part wave?

The photons, the first particle-wave duality to be studied, carried a Poynting field with them everywhere they went. Any interaction could be analyzed in terms of electrons emitting and absorbing photons. The quantum probability of this occurring as the electron moved was called α, and is almost exactly one over 137. This is the ratio of the electric charge of an electron squared to Planck's constant. The beauty of this number has transfixed physicists for nearly a century to the extent that some of them have become unhinged pondering its meaning. Wolfgang Pauli became so distraught over his obsession with the number 137, that he went to see Carl Jung, the great pioneer of Psychiatry.

Pauli's obsession was not groundless, as it turns out. α, or ~1/137 is a key to understanding the universe. We will find α is intimately connected to the number $\sigma = 42.8503...$, which unifies the long range fields of the cosmos.

Paul Ehrenfest was a friend of both Einstein and Neils Bohr, always tried to mediate their differences. He was an accomplished physicist in his own right, having formulated the Eherenfest Theorem relating classical and quantum mechanical systems, and also a proof that charges could find no stable equilibrium in a combination of gravity and electrostatic fields. Erhenfest had many notable students, one was Gunnar Nordstrom, another was Robert Oppenheimer.

Beyond Einstein: GEM Unification

Niels Bohr and Paul Erhenfest, with one of his sons.

Einstein, frustrated by the presence of randomness and probabilities in quantum mechanics debated Bohr at a great physics conference. "God does not play dice with the universe!" Einstein roared. Bohr replied, "Quit telling God what to do."

Bohr and Einstein discussing physics at Erhenfest's home.

Chapter 6. Gravitas: Einstein's Glory

"If you think you have problems with math, mine are greater."
Albert Einstein

Einstein at 14 years old.

Beyond Einstein: GEM Unification

Einstein's youth was an exciting time to be alive and be interested in science. Electricity and magnetism, long treated as two separate forces, had been unified successfully by the brash Scot, James Clerk Maxwell in 1865, even as in America the great war to maintain unification of the United States had been won. This unification theory of Maxwell was bearing rich fruit for science and technology, for not only had electricity and magnetism been unified, but part of this unification was a deep understanding of light, or, more generally, electromagnetic waves. Einstein, gazing at the snowy peaks spent many a warm afternoon imagining experiments with this new electromagnetic wave.

Einstein, by nature a rebel and a dreamer, was an indifferent student at the University of Lucerne. Minkowski, a classmate, would call him a "lazy dog," a statement all the more ironic given the later confluence of their work. Here the influence of Tesla on Einstein's life was once again seen. Einstein, a dashing young man, was much distracted by the affections of his girlfriend Mina Broz, a Bosnian Jew, and a fellow student of physics. She had been a friend of Tesla's, and from that acquaintance knew a genius when she saw one. Given this distraction, and his own rebelliousness in a world that richly rewarded obedience, he did not do well in school. It was unsurprising to outside observers that Einstein, having gained his doctorate in physics, could find no position at a university, and instead was retained only as an unpaid lecturer.

Beyond Einstein: GEM Unification

Einstein and his bride, Mileva Maric.

There might his story have ended in failure, a new Ph. D. without a professorship, without salary, and with a pregnant girlfriend. However, Einstein was as resourceful as he was rebellious. He found work in the Swiss patent office, married his sweetheart, and was soon gazing at the alpine peaks out of the windows of a nice office in Bern. Across this desk poured a parade of ingenious Swiss inventions and ideas that he had to analyze and approve or find flawed. Most of these were electrical inventions, for electricity, through the light bulb and through the electric generator and motor combination, was transforming the world Einstein knew. Einstein, from the moment of his birth, grew up in the age of Tesla.

Beyond Einstein: GEM Unification

The work of Tesla had by this time revolutionized industry and everyday life. Tesla's induction motor, his poly-phase alternating current, and the transformers designed by Steinmetz were now found everywhere. Tesla's ideas were also to be found everywhere. His concepts of electrically powered flight, wireless communications, and power transmission filled the world and appeared everywhere in the stream of proposed electrical patents moving across Einstein's desk. This was not grand physics, but it was intellectual stimulation of the highest order. Tesla felt the world would be dominated by electricity and magnetism in the future, and Einstein, from his unique position as a physicist in a patent office, saw this clearly. He, Einstein, would help make that bright electromagnetic future. From this solid desk, Einstein plotted his next great move in physics. He would assault the summits of physics and dispute there with its great masters.

He organized some friends with deep interests in physics, and met with them to discuss the great news of physics, of which there was much. He gave this fellowship of outsiders the grand name, "The Olympic Institute." Here, and with his physicist wife at home, he sharpened his physical insights with the conversation of other brilliant minds, until he was ready to sally forth with his bold initiatives. He was happy as a junior patent examiner, but in his heart he felt wronged by the physics establishment. He was a vigorous, dashing young man, and brilliant, full of boundless energy and ambition. To such a man, being a government clerk was only a temporary expedient. Einstein knew he was not only the best physicist alive on the planet, but that he would soon prove it. He would return to the field of battle of physics, and regain his honor as a paid professor by besting its champions. In his Olympic Fortress, immune from squabbles over faculty privileges, and the frowns of deans and department chairmen, he prepared his great missives to physics.

The Olympic Institute of Bern: Albert Einstein with friends Habicht and Solovine in 1903.

It was a time of great tumult in physics. Max Planck of Germany had discovered that light, an electromagnetic wave field, seemingly a continuum like space itself, was none the less organized into "quanta" or discontinuous packets. It seemed that light interacted with atoms of matter, in an atomic way. Light, despite being a wave, was also a particle. Using this idea Planck had explained the rainbow of radiation from heated bodies, red when hot, yellow white when hotter, then bluish white like a star, when hotter still. His theory predicted precisely the spectrum of electromagnetic radiation emitted by a black object at a given temperature. It was unmistakably true. The aftershocks of this discovery were shaking physics to its core. Yet the

Beyond Einstein: GEM Unification

world of physics was deeply grateful to Planck. Despite all the trouble his idea was engendering, he had saved physics from a catastrophe.

The ultraviolet catastrophe, that great dragon that Max Planck, a recognized knight of the realm, had slain with his quantum lance, was a crisis of physics. Physics encounters such intellectual catastrophes periodically. One confronts the edge of the map of one's scientific knowledge and writes in the blank space beyond, "here be dragons." These crises occur when one extends what one knows logically, and encounters nonsense. It creates great unease, but what is needed then is a brave champion to sally forth and slay the dragons, to push outward the checkered map of the known, into the circle of the chaos that bounds it. So physics proceeds by a series of successive approximations, each one more exact. Slowly the unknowns are squeezed into the noise of ever refining measurement, and the spurious results of yesterday lead the breakthroughs of tomorrow. The complete and elegant theory of today is soon the cobbled together patchwork of a working hypothesis as new phenomena are discovered in the laboratory or the heavens. The physics we do today is merely the first approximation to what we will do in the next millennium.

Einstein sallied forth in 1905, not with one bold initiative, but with three. It was a measure of his confidence, bordering on audacity, that he did not limit himself to one line of attack. He saw several things clearly that others saw as imponderable mysteries. Science was rich in mysteries then. Technology had advanced to allow measurement of things heretofore considered immeasurable: Atoms had positive cores and negative mantles, the atomic nature of light was apparently the atomic nature of energy within the atom itself, radium put forth a fantastic amount of energy per weight compared to burning coal, and supposedly, could cure

Beyond Einstein: GEM Unification

hoarseness as well. Einstein saw many targets to aim for, including the photoelectric effect, the fact that the particles of light found by Planck, photons, strike electrons like billiard balls. This effect makes solar power possible. Einstein also analyzed the dance of dust particles in air or water, which he showed was the indirect evidence of the motion of atoms. Finally Einstein proposed relativity, the door to a new age. Any one of these ideas, if embraced by the physics community, would be his ticket out of the patent office and into a professorship. 1905 would be Einstein's "miracle year," for he hit every target. Soon, he was a paid professor at the University of Zurich.

Relative to his previous job in the patent office, his professorship was a step upwards, but Einstein was still the passionate young prince he had been before. In his new position his boldness increased. He published bold new theories on quantum theory and the atomic theory of matter. He found that his theory of relativity predicted that mass could be made into vast amounts of energy, leading to the atomic age. He studied the interaction of high energy atoms and light and invented the theory of the laser. He corrected the work of Planck and found that the quantum theory of the electromagnetic field meant that we were surrounded by a sea of electromagnetic energy, the "ground state energy of the vacuum" as it was called then. It seemed a minor detail at the time, but Einstein had discovered the existence of the ZPF (Zero Point Fluctuation), as it was later called. It would be one of his very greatest discoveries and full of meaning and portent unguessed at that time.

Max Planck arrived in Zurich in 1913, soon to have his own Nobel Prize for physics (he received it in 1918). Germany had unified itself in the mid 1800s under Bismarck and had become the scientific and industrial powerhouse of Europe. The heat of fusion of the many German kingdoms

Beyond Einstein: GEM Unification

under the central control of Prussia had made Berlin an incandescent capitol of physics and technology. The Kaiser Wilhelm institute was formed to bring the flower of Germanic science under one roof so they might all shine more brilliantly. Einstein was of German blood, therefore Max Planck had come to recruit this 'new champion of physics' to join his constellation of stars. Einstein accepted after much thought. It was a fine thing to go from the sitting room of the Olympic Institute to the stone halls of the Kaiser Wilhelm in ten short years, to go in one decade from an unpaid lecturer to one in a blazing constellation of stars in the northern heavens. Going to Berlin was a fateful decision to, for Einstein was not just German, he was Jewish. His trip to Berlin would lead to his greatest living triumph but also his greatest life tragedies.

Einstein, even as he finalized his special theory of relativity, had been thinking of gravity. The special theory of relativity focuses on bodies in uniform straight line motion. In general relativity, however, Einstein was to expand this analysis to bodies moving in any kind of path, such as curved paths of planets orbiting a star.

Gravity moves everything in the universe in the same way. This was report by Galileo at Pisa. This is because the larger the mass of a body, the larger its inertia, or resistance to acceleration but at the same time the stronger it feels the force of gravity. That is, the mass that serves as the source of gravity and the mass that serves as the source of inertia are the same. This is called the equivalence principle. This means that what we call gravity is the curvature of spacetime that we all experience the same way. For bodies small compared to the body they orbit, it is as if all the small bodies have the same mass. This "all bodies move equally" was reduced to a variational or economic principle by Einstein. Einstein postulated that spacetime was curved and all bodies moved on the shortest possible path through it. He then postulated that matter curved spacetime. For the mathematics of curved spacetime he used the techniques and formalism worked out by Bernhard Riemann some years before.

Bernhard Riemann, who had died in 1866, had been primarily a mathematician, but his real dream was to unify the gravity and electro-magnetic fields using advanced mathematics. He expanded mathematics beyond the four dimensions normally considered, believing that in higher dimensional systems the fields could be unified. Riemann developed the idea that we could live in curved space where even yardsticks were bent. Like our existence on the curved surface of the earth, the curvature could be subtle.

Bernard Riemann, the master of curved spaces.

Beyond Einstein: GEM Unification

We look around us on Earth, and it looks like a flat world that we can subdivide nicely into square fields for farming. But what we live on is actually a sphere of enormous radius compared to our size. Only by careful observation and thought, such as by the Greek Eratosthenes, who measured the size of the Earth with shadows along the Nile, can we determine that we live on a curved spacetime. What we regard as an orthogonal coordinate system, dimensions we think are totally "right" with respect to each other: length, breadth in the horizontal plane, and height in the vertical plane, are actually not quite orthogonal. What we regard as vertical is in fact a line that extends to the sky above in one direction but to the center of Earth in the other direction. What we regard as two directions in the horizontal plane, far east and far west, maximum north and maximum south, actually wrap around Earth and become the same place. So Riemann considered that we could live in a curved spacetime and not perceive it directly, but instead discover it only through careful mathematical measures.

One sort of Riemann measure is to go out on the ocean and sail one's boat in circles to measure π. If the ocean was perfectly flat and one measured the circular voyage carefully, one would find the circumference of the circle to be always π times the diameter of the circle one sailed. However, the ocean is not flat, so the circumference of your path is always short of π, and, in fact, if one sails in a wide enough circle, the circumference of the circle is finally the circumference of Earth itself and can become no larger, and the diameter of your path, based also on measurement along the ocean surface, is half the circumference of your voyage, so your measurement says π is two.

Instead of looking at the globe from afar, or sailing on the ocean, Riemann designed mathematical measures to be done locally so that one could, in effect, measure the size of Earth

Beyond Einstein: GEM Unification

by walking around the block and measuring each step with mathematical precision.

To Einstein, Riemann's work represented a treasure trove of mathematical tools for expressing his concepts of general relativity. To describe how everything moved he had to describe the spacetime everything experienced.

After some fits and starts, Einstein finally published his general theory of relativity in 1916. He had been greatly helped in his research by David Hilbert, a brilliant German mathematician many years Einstein's senior. Hilbert obtained Einstein's field equations for gravity from a variational principle. In Hilbert's formulation, curved spacetime behaved as if the presence of curvature stored energy. Hilbert worked out a mathematical formulation where Einstein's field equations resulted from spacetime relaxing to a state of minimum energy. To Hilbert spacetime was like a big block of rubber that wanted to wiggle around until tension within the rubber was evenly distributed everywhere.

An everyday example of the power of variational principles is blowing a large soap bubble on a summer afternoon. If one uses a wire coat hanger to make the bubbles and a garbage can lid full of soapy water to dip the hanger, one can blow beautiful bubbles of enormous size. The bubbles break off the hanger in a mild breeze and form long sausage-like shapes, which then lazily relax into spheres while preserving the volume of air enclosed inside. In doing this they are minimizing "surface energy," the source of surface tension. The end result of this lazy relaxation, the sphere, is the object of minimum surface area for a given volume. That last result is mathematical, but the physics principle that drives it to occur is surface tension, or the energy per unit area of the soap film. Therefore, we see from this that mathematics is the description of what physics does, and in order to understand the process we must assign real energy to the mathematical quantities, and real forces to the variations of those quantities.

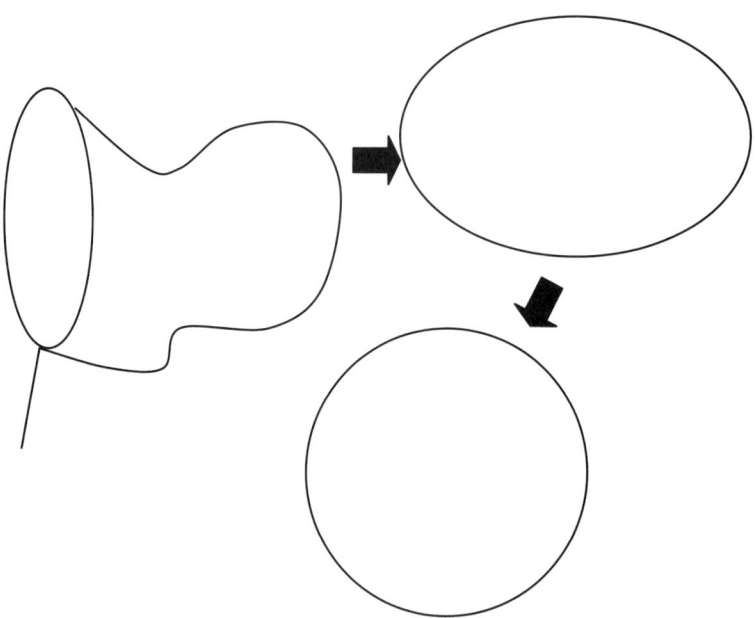

Formation and relaxation of a soap bubble to a spherical minimum energy state.

The Hilbert action principle says that in the four space of spacetime, the curvature generates energy, as if spacetime was a block of rubber that was being twisted and deformed. The equation that results from Hilbert's action principle is an equation that distributes the stress throughout the rubber block of spacetime so the total energy of stress in the rubber is a minimum.

Beyond Einstein: GEM Unification

The new theory of Einstein went far beyond his special theory of relativity in scope and meaning. The special theory said that mass could be converted to vast amounts of energy and that time was variable like space, but the conditions where these things occurred were far from everyday experience, like moving at near the speed of light. The general theory of relativity however, dealt with where we lived, in the curved space of a gravity field.

The general theory of relativity was based on two postulates: One, that matter curved spacetime. Two, that all particles followed the shortest possible path though space time. Particles would move in straight lines in space without curvature, but in curved space their paths would be curved in the effect we call gravity. Gravity arose through the curvature of space by mass.

Gravity was not just an acceleration, it was the local curvature of space. To ride in a closed elevator being pulled at constant acceleration was to experience a feeling of weight like gravity, but one could always tell this from real gravity. Real gravity, according to Einstein, curved space. One could test the curvature of space by a Riemann test, in the case of the elevator by releasing a circle-shaped pattern of ball bearings and watching them fall carefully. The circle would be formed by ball bearings in the gravity field with the circle extending vertically. In a simple acceleration in the closed elevator, the pattern of ball bearings would simply fall together and the circular pattern of bearings would remain unchanged. However, in real gravity, the circular formation of ball bearings would distort as it fell, the circular elongating in the vertical and compressing in the horizontal. This is the same pattern that the tidal distortion of Earth and its seas experience due to the gravity of the Moon and Sun. The distortion of the circular formation, this tidal distortion pattern as the ball bearings fall, tells us that we are in a true

Beyond Einstein: GEM Unification

gravity field, that we are in a region of space that is curved by a nearby mass. However, in a region outside the mass, if we ride in a space station and place a lead weight in the center of a chamber that is weightless and put the same diamond pattern around it we get a different pattern, they all move inward towards the lead weight. This pattern tells us we are in a region of curved space due to mass density. In keeping with Riemann's concepts, these tests of spacetime curvature are local tests rather than "big picture" global tests.

Gravity can be mimicked by acceleration, as seen in an accelerating rocket .

Beyond Einstein: GEM Unification

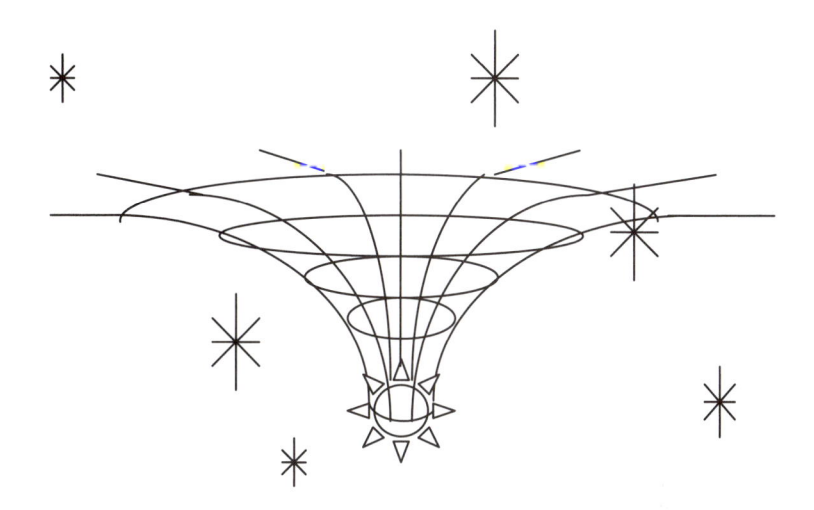

Curved spacetime around a star.

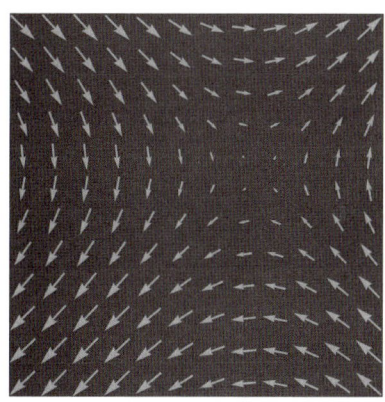

A tidal distortion vector field.

Beyond Einstein: GEM Unification

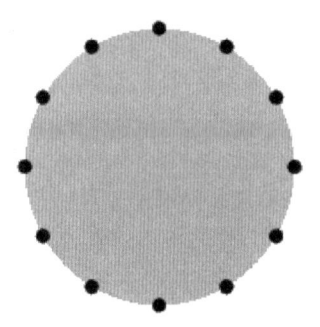

A circular pattern of ball bearings falling in an accelerating rocket.

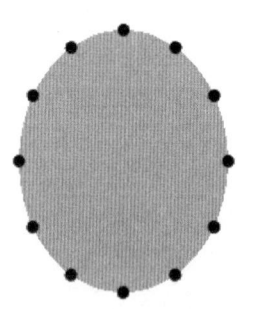

A circular pattern of ball bearings falling at the Earth's surface and being distorted by the tidal curvature of spacetime.

Beyond Einstein: GEM Unification

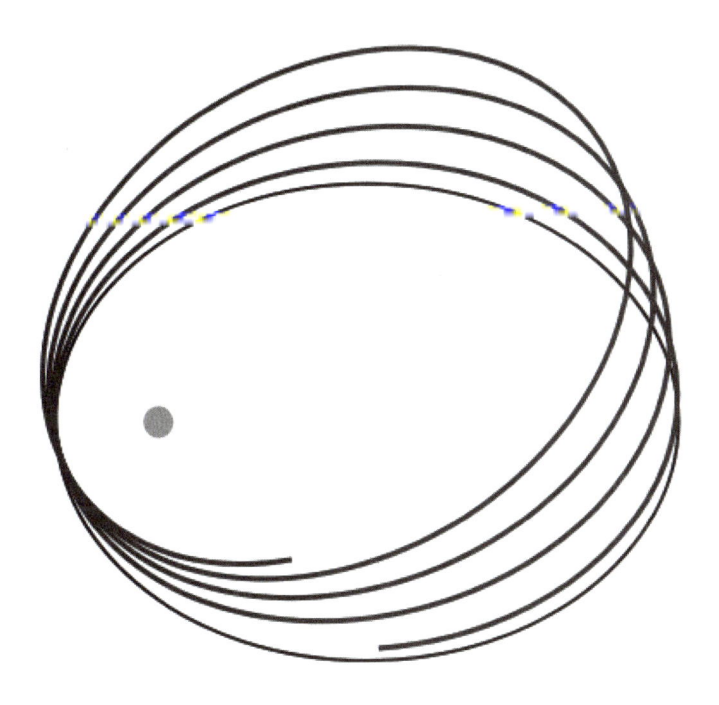

The orbit of Mercury around the Sun due to general relativity.

The second postulate, the postulate of how particles move in curved spacetime, was based on an economic concept similar to the Hilbert action principle. The shortest path between two points on a flat piece of paper is a straight line. However, the shortest path on a curved surface is itself a curve. On Earth's surface, the shortest path between two points is a section of a "great circle." On a flat map of the world this looks like a curve for the simple reason that the map is actually a distorted representation of the curved surface of Earth. In later formulations of quantum mechanics, it became apparent that the wave field of matter ψ was constantly seeking the shortest path though spacetime and those parts that found it moved forward more quickly, so the whole wave packet of matter sought the shortest path automatically, as if it was racing itself.

These ideas, that spacetime minimized its overall curvature caused by mass, and the curvature of space is visualized by the paths of particles that were themselves minimized in length, are profound but seem fairly benign. They suggest an optimum cosmos, where, as Aristotle described gravity, "everything seeks its place," and all is right with the world. However, the reality of Einstein's general theory was that it did away with "everything being right"; "rightness" no longer existed. Everything, including light rays, was now curved. When Newton put forth his theory of gravity, it gave order and a sense of stability to the heavens; when Einstein produced his theory the heavens would become warped and nightmarish entities would become their inhabitants.

Mercury, the closest planet to the Sun, had long been known to move in an elliptical, that is, oval-shaped orbit. Newton's theory of gravity predicted that any deviation from this perfect orbit was due to the influence of its neighbor Venus periodically moving near it. However, the ellipse of Mercury's orbit, so fast and with enough oval shape to be easily measured, was rotating in time. Instead of following a perfect oval path it was tracing out a daisy shape that would not repeat from one orbit to another. Einstein showed this was the subtle yet irresistible consequence of the curvature of space near the Sun. Gone were the beautiful ellipses of Newton in flat space; now space was warped and the perfect retracing orbits were gone, replaced with daisy-like spirals. The eternal orbits of Newton were now replaced with orbits that decayed in time, because the orbiting planets generated gravity waves and, therefore, lost energy slowly over time.

Beyond Einstein: GEM Unification

But the worst was yet to come. Einstein's theory produced an equation for the evolution of the whole cosmos, and it projected its doom in either death by fire or death by ice. To see why this is, one must consider a simple spherical universe with stars and galaxies arranged in shells around a center. The cosmos is empty space studded with stars. The stars all attract one another by gravity, so the cosmos cannot be at rest. The stars and galaxies all will fall towards each other and finally collide. This is the death of the cosmos by fire.

Alternately, the spherical cosmos can be expanding outwards as if born in a huge explosion with every galaxy moving away from every other one. If it expands rapidly enough it will overcome its gravity and expand forever. But then the stars will all use up their hydrogen and become balls of ashes, grow cold, and the cosmos will die the death of ice as even the stars all cool and grow dark. The cosmos Einstein created with his theory was an unsettled, disturbed, warped cosmos, not like Newton's great beautiful eternal clockwork. But the worst thing he had introduced, a nightmare of physics, was yet to be discovered in his theory.

A German professor of physics, Karl Schwartzchild (transliterated "black child"), moved by patriotic fervor, had joined the army at the outbreak of World War I in 1914. He had served in the artillery on the Russian Front. To escape the horrors and boredom of that front he had managed to produce an exact solution to the complex Einstein equations. The solution was nightmarish, however. It said that if a star shrank too much it would curve space to its breaking point around it and it would cut itself off beyond all escape. Not even light could escape the surface of the star and the star would become a bottomless pit. Black-child had discovered the Black Hole. He published this result, and Einstein was amazed by it. But it was a last burst of brilliance for

Beyond Einstein: GEM Unification

Schwartzchild. He himself contracted a terrible skin disease shortly thereafter because of the horrid conditions on the Russian front. He returned to Germany to die.

The war was claiming more and more of the life of Germany. Max Planck's oldest son was killed in the trenches at Verdun. The British naval embargo was reducing Germany to starvation. It became obvious, with American entrance into the war on the side of French and British, that Germany's war was lost. Einstein, always a critic of militarism and unquestioned authority, signed a petition, along with many other university professors, calling for peace. Finally, in 1918, after the slaughter of millions of the bravest and brightest of Europe's young men, the guns fell silent. However, not everyone saw the end of the war as a triumph of human intelligence over homicidal mania. An Austrian corporal named Adolph Hitler wept openly in sorrow.

An English astronomer named Eddington became entranced by Einstein's theory and mounted an expedition of an island off Africa to witness a great solar eclipse. Miraculously the heavy clouds seen that day parted just in time for the eclipse to be photographed. The eclipse had been magnificent in its totality, so much so that stars had been clearly visible in the sky around the darkened Sun. Careful measurements of the stars' positions near the Sun showed that their light had been bent by the Sun's gravity. Einstein's theory was verified, and he instantly became the most famous physicist on Earth. However, the very same trait that made him acceptable to the world as a German physicist, his opposition to the war, excited bottomless hatred among the nationalistic segment of the German population. They felt that Einstein and his other professors, many of them Jewish, had "stabbed Germany in the back"

Beyond Einstein: GEM Unification

during the war. Einstein had become a man of Gravitas, a man of consequence, but also bearing great weight.

Gravity bending light around a massive galaxy in the center.

Eddington's student from India, Subrahmanyan Chandrasekhar, encouraged by the success of Einstein's theory, made a bold hypothesis concerning the death of large stars. The pressure exerted by a gas is due to the microscopic motions of atoms bouncing off each other and the surfaces that confine them. Chandrasekhar used the fact that the motion of atoms that created thermodynamic pressure increased the mass of the atoms slightly, thus increasing their mass through special relativity. This meant that gas at very high pressures would weigh more than the same volume of gas at lower pressure. Chandrasekar extended this idea to its logical conclusion: For a star above the "Chandrasekar Limit" in mass, with no more hydrogen or other light elements left to burn to create fusion energy, the collapse of the star could not be stopped. It would collapse forever.

For a star with no new energy source, the only thing opposing its gravity was thermodynamic pressure, but the higher the gravity the higher the pressure, so that a star's weight would increase due to the pressure increase itself. The increased weight of the star due to its own pressure would in turn create more gravity, generating still more internal pressure. In a nightmarish cycle of events gravity would feed on gravity and pressure upon pressure as the star crushed itself. Finally, the star, once much larger than the Sun, and more brilliant, would shrink to the size of Earth, then to the size of the moon. Reaching unprecedented temperatures it would shine even more brilliantly in a last desperate flare of light. It would continue to crush itself to a few kilometers in radius and there the Schwartzchild solution of gravity would hideously emerge, creating an "event horizon" that not even light could escape. The star's once brilliant light would be snuffed out by the force of its own monstrous gravity. Inside the spherical event horizon the star would now collapse in on itself to the size of an orange and finally to the size of an electron. It had become a black hole.

The collective mind of the physics community recoiled from this hideous possibility. The concept of a star radiating light for the whole cosmos strangling itself by its own gravity, then turning into an abyss of darkness, was beyond acceptance. To embrace it would be to concede that darkness could triumph over light. It was to accept that the brightest star in the sky could degenerate into a hideous black nightmare of twisted spacetime, impossible to escape. The physics community rejected Chandrasekhar and his nightmare, and with his theory he himself was rejected by his mentor. Chandrasekhar managed to go to America and find a new position, but the price he paid for telling the world what it did not want to hear was deep. It would take many years for his reputation to recover.

Subrahmanyan Chandrasekhar, the father of the Black Hole, the ultimate catastrophe of physics.

Einstein managed to enjoy his newfound fame. He was now divorced from his first wife and so married his cousin. He was laboring mightily now on what he believed would be the crowning glory of his career, the unification of gravity and electromagnetism. His unification theory attracted much interest in those days, but alas, the problem was much more difficult than he had anticipated. He made no progress. Because of his fame and open mindedness, however, he found a new role in the physics community as a last court of appeal for new ideas that were having difficulty being published. In this role he served as a patent clerk again for new ideas. His most important contributions in this role were in the then burgeoning world of quantum mechanics. However, in 1921 Einstein endorsed a revolutionary paper for publication. He did not like the paper but its results were remarkable. This was the idea of a hidden fifth dimension proposed by Theodore Kaluza. With this fifth dimension, Kaluza had found that both the equations for gravity and EM, the Einstein's equations and the Maxwell's equations could be found in one mathematical procedure. Encouraged by this article, Oskar Klein of Sweden found that the theory worked best if the fifth dimension was hidden because it was curled up on itself. The Kaluza-Klein theory of a curled up fifth dimension was remarkable because it was so simple.

The Hilbert action principle led to Einstein's equation for gravity by reducing spacetime curvature stress everywhere in a four-dimensional spacetime. Kaluza and Klein had merely added a fifth dimension that was "compact" and had redone the Hilbert action principle in a five-imensional space rather than four. In five dimensions gravity occurred as before, but another set of equations had magically appeared, Maxwell's equations. It was astonishing. Wolfgang Pauli, a brilliant naysayer, dismissed the whole procedure as "mathematical sleight of hand." However, it looked to many in the physics community that Einstein's goal of a unified field theory was near to completion. If the mathematics of EM and gravity could be unified in one expanded Hilbert action principle, could not the physics be unified? However, Einstein, in a rare instance of agreement with Pauli, rejected the idea of a hidden fifth dimension. Einstein was, at his heart, despite his rebellious intellect, a conservative in certain ways. To have such important physics depend on a hidden dimension seemed to him inconsistent with the beauty of physics as he had experienced it. Einstein endorsed the article, but rejected its basis. Einstein was having enough trouble with the mysteries of quantum mechanics, which injected statistics into his perfect deterministic vision of physics. It seemed fundamentally unphysical to him. Quantum worked, but its premises and concepts were wrong,

Beyond Einstein: GEM Unification

Einstein felt. He continued to protest loudly at conferences, that, "God does not play dice!"

Gravity waves from two black holes orbiting each other (NASA).

Beyond Einstein: GEM Unification

Chapter 7. The Aurora

"All I want to know are God's thoughts when he made the universe; the rest is details."
A. Einstein

The night sky is full of lights: the stars, the aurora, the lightning. All of these are due to the plasma state of matter. Plasma is the fourth element of Aristotle, fire. However, it is much more than fire, it is a gas that gives off light and is much hotter than the air around it. Plasma is the trueflame, the fire of the Sun and stars, of which the fire we make on Earth is but a shadow. Plasma is a gas in which the electrons and positively charged nuclei of the atoms have come apart and now comingle as two fluid entities, one negative and one positive. This is the essence of 99% of the visible matter in the universe, which is bound up in stars.

Beyond Einstein: GEM Unification

The Aurora.

Plasmas act like a dynamic entity; the charges within them move at the slightest electric field, as in metal, and conduct electricity. They pulse with electric and magnetic waves. Magnetism confines a plasma and can shape it. The plasma often finds itself flowing along the magnetic lines of force, but seldom across them. The hotter the plasma the better it follows the magnetic lines of force.

One sees another level of richness when one combines the effects of electric field and magnetic field onto the particles of a plasma. If one makes a thin plasma of hydrogen between two plates in a vacuum chamber and places different voltages on the plates the electrons will fly one way and the protons the other because of their different charges. The charges will move faster and faster as they flow past each other in this thin plasma, each towards the plate charged opposite to each other. Now, however, we turn off the voltage and restore the plasma to relative stillness and instead impose a magnetic field parallel to the faces of the plates. The Lorentz force created by the motion of the electrons and protons combined with the magnetic field will cause the electrons and protons to move in circles around the magnetic lines of force. So we have seen in this simple experiment motions of the particles in the plasma when electric or magnetic fields are present. Now we will do a new and remarkable thing.

Earth's Magnetosphere.

If we recreate the electric field and add to it the magnetic field a marvelous thing occurs. We create a new field for one thing, and the combined magnetic and electric fields create a Poynting field at each point, but they do something else marvelous. The electron and protons of our simple hydrogen plasma do not accelerate in the electric field and they no longer move in circles, but instead they move together. The protons and electrons both assume the same speed and move in the direction of the Poynting field. This effect is used in many devices using plasmas, especially a device called the cyclotron. The velocity along the Poynting vector field is called the "ExB drift" or Poynting drift, and it is remarkable in that it affects all charged particles the same, regardless of their charge and regardless of their mass. We can do a further trick in this simple experiment; we can tilt the plates around an axis of magnetic lines of force so they are closer together at the bottom than at the top. This makes the Poynting field stronger at the bottom of the plates than at the

Beyond Einstein: GEM Unification

top. The protons and electrons do something equally marvelous; they accelerate, moving faster and faster along the Poynting field, going faster as it becomes stronger. This is a simple experiment, performed in any well-equipped University laboratory, but, as we shall see, it contains the world.

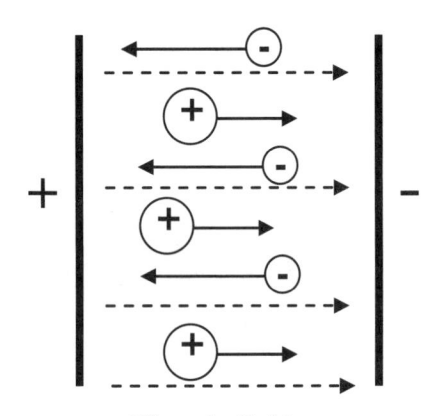

The motion of protons (+) and electrons (-) in a thin plasma exposed to an electric field.

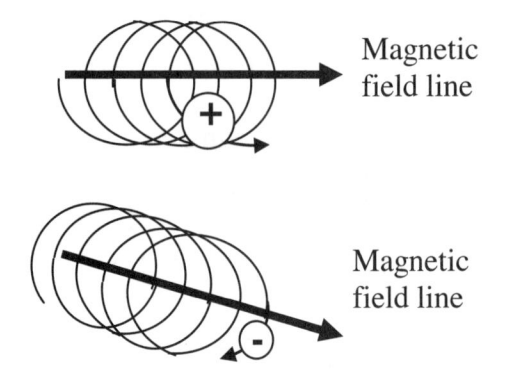

The motion of electrons and protons in a hydrogen plasma with a magnetic field.

Beyond Einstein: GEM Unification

Langmuir investigated the plasma state in the early 1920s and discovered many of its mysteries. He found that electric probes, consisting of a stub of bare wire, would give far more information about a plasma than would any other test. Night after night he would imprison a thin gas in a sealed gas vessel at a millionth of an atmosphere of pressure and then apply voltage to it. The gas would ignite into a glowing scintillating flame inside the glass, following its shape like a liquid. Different gases made every different color of the rainbow. Make the glass tube long and thin and you could bend it like spaghetti and make a neon sign out of it. Langmuir named this true fire "plasma" from the same Greek root word for plastic, as something that could be shaped. Plasmas form wherever energy density exceeds the density of matter by a fixed amount, as in high temperatures or in the presence of strong electricity. Hence, when electric energy density exceeds matter density a spark of plasma forms. Plasmas have been the indispensible tool of science since the late 18th century.

Beyond Einstein: GEM Unification

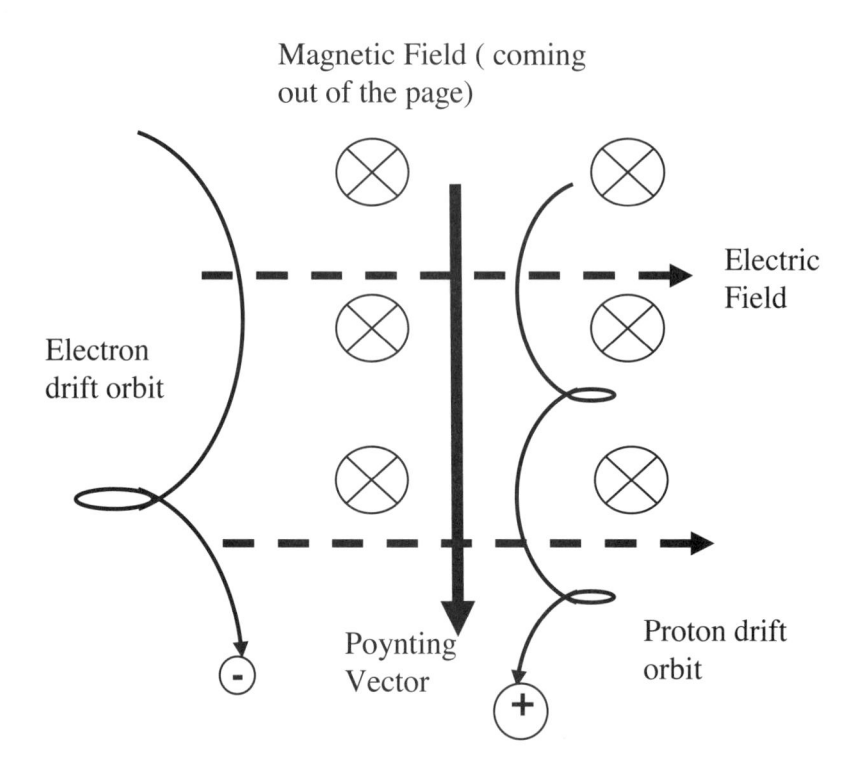

Figure labels: Magnetic Field (coming out of the page); Electric Field; Electron drift orbit; Poynting Vector; Proton drift orbit

The ExB or Poynting drift of charged particles in crossed electric and magnetic fields.

Only in a plasma can one see the dynamics of atoms and light in all its glory; there the atoms are bare and interact with the EM fields of light without disturbance from their neighbors. It was in plasmas of hydrogen, that the first line spectrum of hydrogen atoms was seen and studied. It was seen that the pattern of lines of perfect color the hydrogen emitted followed an orderly pattern. This study would later lead to the discovery of the quantum nature of matter. However, because plasmas form wherever energy densities are high, plasma has been the handmaiden of power since the beginning of the twentieth century.

The plasma is the sign of nature's power. The plasma of the Sun's surface is shaped by magnetic fields into arches of lines of light following the lines of force. The magnetic arches twist and then release their magnetic energy into the plasma heating it to millions of degrees and sending it bursting like the waves of an angry sea out into space. It crosses the empty spaces to earth and crashes into the lines of force of Earth's magnetic field. The dynamics of this collision out in space creates bolts of electricity to flow down the lines of force of Earth to grace the polar skies with aurora. Deeper in the atmosphere the dynamics of the change of state of water from gas to ice in clouds creates massive electric fields that find release as lighting bolts of plasma between the clouds and down to Earth. The ancient Greeks, knowing the ruin that could result from getting in the way of such a process, made the lighting bolt the symbol of the power of their most powerful god, Zeus, called by the Romans Jupiter. So it was that plasmas became the sign of power for governments.

Beyond Einstein: GEM Unification

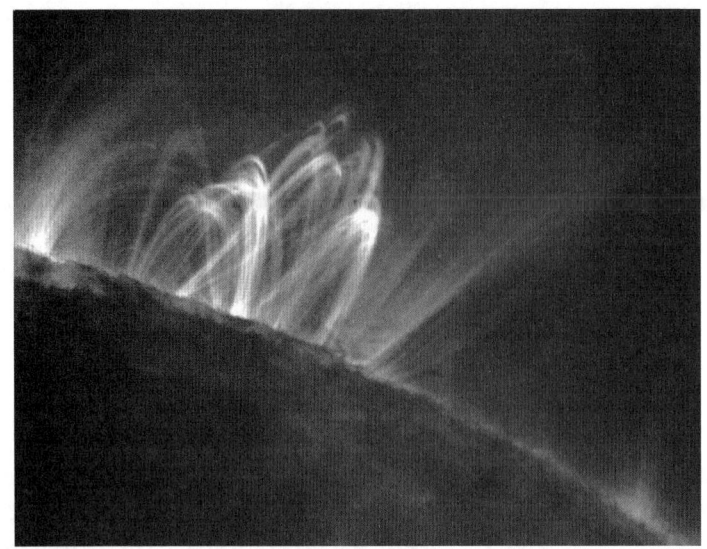

Arcs of plasma following magnetic lines of force on the Sun.

To study atoms and their substance plasma machines called cyclotrons were made, and with these it was found that atoms of matter could be separated into isotopes by electric and magnetic fields. Some isotopes were found to be more valuable than others. In 1934 it was found that an isotope of uranium could unleash unheard of destructive power to whatever government could isolate it in quantity. In Oak Ridge Tennessee, the power of several hydroelectric dams was harnessed to run a whole building full of cyclotrons. The uranium 235 they separated would be dropped later on Hiroshima. These were terrible days.

The terrible vision of Chandrasekar that had so horrified physicists and astronomers a decade before made a new apparition. The prospect of a brilliant star shining of plasma in the heavens turning into a nightmare of darkness and twisted spacetime, so unthinkable in 1927 when it was introduced, was now seen as the inevitable and inescapable

Beyond Einstein: GEM Unification

fate of massive stars whose hydrogen was exhausted. They would indeed begin a process of collapse upon themselves, crushing themselves out of existence. As they collapsed they would begin to spin more rapidly, like an ice skater twirling and pulling her arms tightly inward to her body. Alas, the centrifugal force this faster spinning generated did no good against the irresistible forces of gravity pulling the star inward. The new frantic spin only increased the energy of the star now, making it even heavier. The light of the star got heavier as it collapsed, increasing gravity even more. Finally, the terrible equation of Schwartzchild emerged to replace the gravity of Newton and the surface of the once brilliant star became an event horizon of blackness, from which nothing, not the waves of bright plasma, not even light itself, could escape.

Unlike the imagined ultraviolet catastrophe of theoretical physics before, this catastrophe was real and had no remedy. The man who saw this clear as crystal was J. Robert Oppenheimer, a Jewish American son of wealth, who had seen for himself Europe metamorphose. Oppenheimer published his great work of physics in July 1939, showing the terrible vision of Chandrasekar to be true. The next month Hitler invaded Poland.

Beyond Einstein: GEM Unification

J. Robert Oppenheimer, the father of the atomic bomb.

The bright star of Europe that was Germany, now driven by history, driven by geography, and driven by the urging of one hypnotic devil in human form, Adolf Hitler, was collapsing on itself. To be Jewish was to become a marked person. To be an intelligent person in Europe was to become stricken with fear and loathing.

Paul Erhenfest and a son with Einstein.

Beyond Einstein: GEM Unification

The event horizon around Nazi Germany was forming, and the last atoms of light tried to escape. When Hitler came to power in 1934 Einstein decided to visit America with his family and secretary and not return. His friend Paul Ehrenfest in nearby Belgium also saw things clearly, but used a different escape: he took his mentally handicapped son to a park and shot him with a revolver, then blew his own brains out. Since 1938, when Hahn and Strassmann split the uranium atom with neutrons, and Lise Meitner had shown that this would release both energy and more neutrons, the possibility of chain reaction and a nuclear weapon was realized. Wiser minds knew that the first person to gain the power of this weapon stood to be Adolf Hitler. One of those minds was Ehernfests's student, Robert Oppenheimer, another was Einsteins' student, Louis Szilard.

Leo Szilard, an old friend of Einstein's, was one of the first men to see the terrible possibility of Nazi Germany gaining an atomic bomb, and composed a letter to President Roosevelt, warning him of this possibility, with another friend Edward Teller. Together Szilard and Teller went to Einstein with it. Einstein signed it, Roosevelt read it, and a whole chain of powerful events was begun. Einstein had saved the world, and also unleashed a new terror into it. Soon even as the lightning of war raged over Europe and the Pacific, in a great hall plasma shimmered and flowed, separating the atomic wheat from the chaff and condensing the atoms of uranium 235 into little pieces the size of the head of a pin. My uncle, William Siri, a physicist from Berkeley, would gaze into thousands of miniature cyclotrons, optimized for this task, and watch the plasma dance with gossamer light, and he would wonder at this plasma.

Beyond Einstein: GEM Unification

Leo Szilard, who saw things clearly.

Edward Teller, as Director of the Lawrence Livermore National Laboratory.

Beyond Einstein: GEM Unification

In Russia, laboring in the mighty war plants that strained to supply the Red Army with ammunition, was Andrei Sakharov, watching the aurora of the long Russian nights. He had shown himself to be an able problem solver in the mighty effort to save Russia from the Nazis, and soon he was promoted to making armor piercing tungsten cores. Sakharov found himself near the end of the war laboring with the finest minds of Russia, Landau, Kurchatov, and Zeldovich on a new problem.

At Alamagordo, New Mexico on July 25 in 1945 the night-time desert was lit with the searing light of a million degree ball of plasma as the first nuclear weapon was detonated. For Oppenheimer it was a sad triumph. *"I am become death, the destroyer of worlds,"* he lamented, quoting the Bhagavad-Gita. The enemy Oppenheimer had sought to use it against, Adolph Hitler, was dead and the Nazi war machine was destroyed.

Trinity nuclear test with plasma fireball in the early stages.

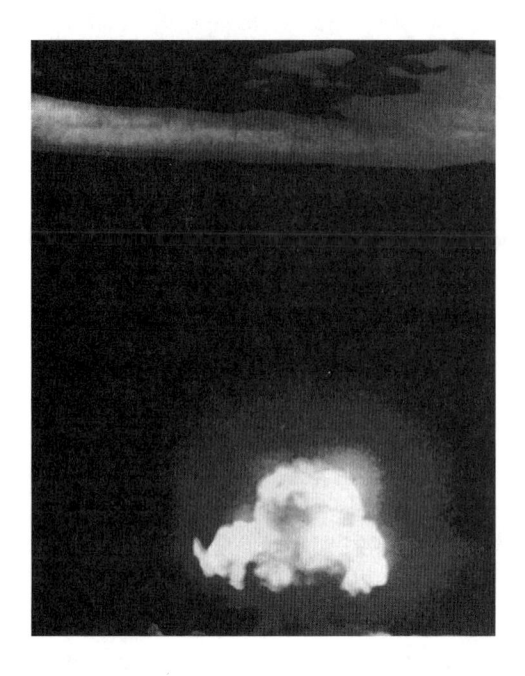

Beyond Einstein: GEM Unification

Trinity nuclear test after plasma fireball has developed.

Oppenheimer's counterpart in Nazi Germany, and old acquaintance, Werner Heisenberg, had failed to harness nuclear energy in Germany, despite their tremendous head start. There was no Nazi atom bomb, despite the best efforts of what remained of German science, for many of its finest minds had been Jewish before the war, and had fled or been murdered. The distractions of a conventional war, and finally the deprivations of the allied bombings had crippled the Nazi bomb effort. However, perhaps the greatest obstacle to the problem lay in the mind of Heisenberg himself, the same mind who had authored the Heisenberg uncertainty principle. Heisenberg was himself uncertain that he wanted a historic mass murderer, Hitler, to have a new and much more powerful means of mass murder at his disposal. So Heisenberg had presided over a vast effort, yet been confused as to how to proceed. Finally, Heisenberg had turned the German bomb effort into a huge and unsuccessful physics experiment.

Not so in Los Alamos, where the finest minds in the world, many of them Jewish refugees from Europe, had labored under Oppenheimer's single-minded governance and been backed by the mightiest industrial power on Earth. Oppenheimer anguished over the use of the bomb on Japan, but he comforted himself that America, a noble nation, had the bomb but was loath to use it. That was enough to ensure peace in the world. Surely the accomplishment of his team of geniuses, the atom bomb, thought Oppenheimer, could not be duplicated anywhere else on Earth, in a thousand years.

In Russia, Sakharov was puzzled. He knew Kurchatov was a genius of high order, but the stream of clever ideas he was directing Sakharov and his colleagues to work on was astonishing. It was like the work of a hundred geniuses. Only later did he discover that Kurchatov was sorting the fruits of a vast network of Soviet espionage, whose nexus was in Los Alamos. Beria, the head of the dreaded Soviet Secret Police, was the chief administrator of the Soviet bomb program, and Joseph Stalin was to be its chief beneficiary. No moral ambivalence was evident in the Russian bomb program, for it was a capital offense. The Soviet bomb program, staffed with the most brilliant minds Russia could produce and resourced on a scale rivaling even the American bomb project, labored tirelessly with Sakharov and its elites. Finally, a smile was brought to Stalin's face when the steppes of Russia erupted with the crack of doom. But for Sakharov, an even greater goal now lay beyond this towering mushroom of fire.

Beyond Einstein: GEM Unification

Andrei Sakharov, the genius.

Fission is the capture of energy from a supernova, the last burst of light given up by a massive dying star before its core collapses into a black hole, but fusion is the source-blood of life in the universe. The universe is born out of hydrogen, the lightest element. The hydrogen fuses in the heart of stars to form helium. This fusion process is driven by the strong force. All other elements come from this process repeated at increasing temperature and diminishing profit until iron is reached. At iron, the maximum nuclear binding energy per nuclear particle is reached. Beyond the brow of this peak of binding energy, nuclei become less stable. Below iron in atomic number, fusion can yield energy; above iron, fission yields energy. Finally, nuclei, like uranium, yield the most energy per fission. However, uranium is rare and hydrogen is abundant. A bomb made with hydrogen, and gaining its power from fusion, could be made much larger than a bomb based on fusion. Such a bomb could rival the supernova.

Therefore, the harnessing of fusion energy became the obsession of Sakharov and his colleagues. It was fruitful obsession, for the harnessing of fusion would provide far more power than fission. Uranium has to be mined but hydrogen was found in every raindrop.

Beyond Einstein: GEM Unification

Not all of Sakharov's zeal was for building a fusion bomb. Russia is cold in the winter and harnessing fusion in power plants would provide Russia with limitless light and heat. One method Sakharov thought would harness fusion was to weave a twisted rope of magnetic lines of force into a ring of magnetism. Each layer of the rope of magnetism would have a different pitch of rotation. In this way the plasma could be full of heat and pressure, like the heart of a star, but not wiggle out. The genius of this approach was that the plasma would choose its own form, shaping its own magnetic field out of currents flowing within itself, rather than having human beings try to force it into some configuration it did not like. Sakharov's concept was called the Tokomak and it would become the chief method for harnessing fusion for power. However, he did not let this distract him from the chief problem at hand. Stalin wanted a hydrogen bomb, and Stalin got what he wanted.

The hydrogen bomb, developed independently by Edward Teller in the United States, in a neck and neck effort, utilizes the pressure generated by light, radiation pressure, to operate. That is, it operates on Poynting flow.

Beyond Einstein: GEM Unification

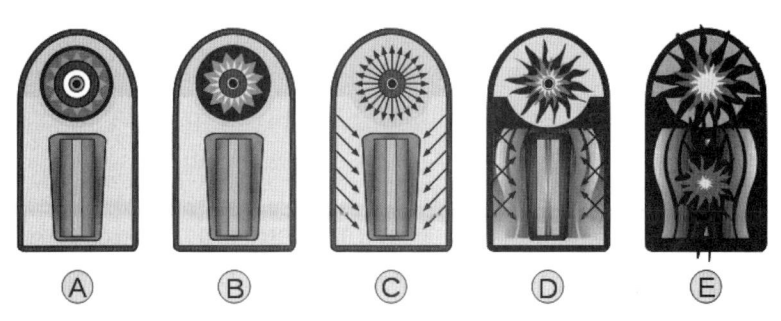

A diagram of the hydrogen bomb and the sequence of its operation. A. The bomb consists of Fission Primary (top) and Fusion Secondary (bottom) in a heavy casing. B. Fission primary detonates. C. Poynting vectors indicate radiant energy flow from the Primary to the outside of the Secondary. D. The Fusion Secondary is compressed by the ablation created by the Poynting flux and implodes, E. The Fusion Secondary detonates.

This idea was once classified, but eventually made itself into public currency. The idea is simple. Objects in a dark box that are themselves bright repel each other with mutual radiation pressure. This force obeys the inverse square law of electromagnetism because it is driven by electromagnetism.

Now, we look at the photographic negative of this image: two dark objects in a bright box. The dark objects are attracted to each other with an inverse square law. They do this because if we stand on either object, we see in the walls a sea of uniform light, like Planck's white hot box of iron from quantum mechanics. However, if another dark object is in the white hot box, one will see it as a black dot in the sea of light. This means that slightly less light, and slightly less Poynting vector, is coming from the direction of the other object than from the opposite direction. The result is that the pressure of the light, the Poynting Force, will drive the two objects together with a force that becomes stronger as they

Beyond Einstein: GEM Unification

approach each other. This attractive force will obey the inverse square law, just like the repulsion of two bright objects in a dark box. In a hydrogen bomb, an atomic bomb turns the box into white hot plasma, and a sphere of hydrogen, viewed as a collection of dark objects, is crushed into a miniature star.

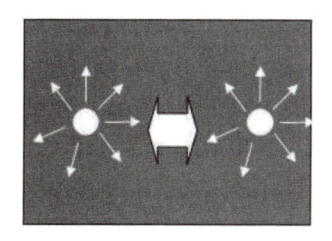

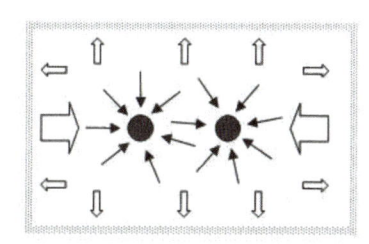

Two Hot Spheres in a Cold Box   Two Cold Spheres in a Hot Box

Sakharov's model of gravity: Two bright spheres in a dark box repel each other due to mutual radiation pressure and two dark spheres in a bright box attract each other due to mutual shadowing.

Sakharov was obsessed with two ideas as he watched aurora in Russia's long winter nights. He had solved the problem of the hydrogen bomb and kept the Cold War powers in military balance; now, he had time to ponder the deep things of physics.

The shimmering lights above him amid the stars reminded Sakharov continually of unearthly things. Had the universe come into being as a shimmering plasma condensing into stars and planets? If the universe had come to be as pure hydrogen, he surmised, the rest of its evolution would follow from laws that operated even now. But what of its beginnings? What laws of physics governed the moment of

Beyond Einstein: GEM Unification

creation, before the everyday laws of physics themselves came into being. The universe had begun as pure hydrogen, protons and electrons, but what had come before? It became obvious to him as he watched the aurora, that what had come before everything was the vacuum itself. This meant, grasped Sakharov, that the laws of physics were different in the moment of creation than we perceived them now. If the universe began as hydrogen and before that vacuum, then the most important law of physics, the conservation of mass-energy had to be suspended. Or did it?

Einstein had corrected Planck's view of the quantum vacuum. Planck viewed a box of absolute zero temperature as having no EM field inside. Einstein had seen this was not so. Einstein showed that the absolute zero state of the vacuum had to be alive with energy. Einstein had discovered the ZPF, the Heisenberg uncertainty fluctuations of the vacuum around zero. The vacuum did not contain a vanishing amount of energy; its energy was an infinity. The energy of empty space dwarfed the mass energy of the particles that moved within it. The void was mightier than the atoms it contained.

How could this be? If we were surrounded by this sea of energy, how was it we could not see it? But if we could see it, it would blind us. In fact, there would be no "us" at all; we would be turned to plasma. The answer, intellectually, at least, was that the energy density of the ZPF was like the pressure felt by fish at the bottom of the ocean. They did not feel it because it was also inside them. Only differences in pressure would make the ZPF perceivable to every-day human life. However, in the world of subatomic particles the ZPF drove them in a frantic dance in the nearly canceling fields. The closer one looked the greater the motion. In German this was called the "*Zitterbewegung*," the quantum jitters. But what then of the infinity?

Beyond Einstein: GEM Unification

Sakharov's colleague Yakov Zeldovich lived in a nearby cottage in the secret "Atomic City" of the Soviet Union where the hydrogen bombs were built. As related to the author by his son, Boris Zeldovich, Sakharov would often visit Zeldovich's cottage on cold winter nights and discuss physics over hot tea. Zeldovich had contrived a way to tame the infinity of the ZPF. The ZPF was a phenomenon of quantum EM theory, but now Einstein's gravity rode to its rescue. Two things happened, Zeldovich reasoned: One, miniature black holes would form in the vacuum of opposite charges, particle and antiparticle pairs, as Dirac had shown, must happen in the vacuum. Two, they would then annihilate; in doing so they would turn the vacuum spacetime into a chaotic foam so no smaller thing could be made. It was as if spacetime was being put through a shredder continually and we formed our existence on the shredded material. The spectrum of the ZPF was then a "cutoff" in EM wavelengths, like the ozone layer cuts off the ultraviolet from reaching Earth's surface. The wavelength at which the cutoff occurs is the Planck Length. The Planck length is, then, the radius where quantum black holes form and disappear by virtue of the Heisenberg uncertainty principle. These black holes are unstable because the event horizon is also the quantum wavelength of the black holes, so that they tunnel out of their own event horizons and annihilate their neighbors. The Planck scale was, then, a scale of maximum turbulence that shredded everything of smaller scale. The infinities were gone. But what replaced them?

Even after Zeldovich's first marvelous scheme the vacuum remained heavier than the cosmos in every square centimeter. This has been called the "most embarrassing number in physics." Therefore, Zeldovich, after first taming

Beyond Einstein: GEM Unification

the quantum infinities with Einstein's gravity, polished off the remaining embarrassments of the vacuum with Einstein's cosmology. Einstein had been unhappy with the theoretical cosmos he had created and he had tried to fix it.

Einstein experienced one of the great embarrassments of his professional life over the cosmos. Einstein, being a sensible person, had been aghast when he found that the theoretical cosmos he had created would end like some Norse legend, dying either a death of fire or ice. The cosmos was atoms and the void, and all the atoms attracted each other. Therefore, a static universe could not exist; all the atoms must collapse on themselves in a fiery heap or be scattered with such force that the universe would expand forever. To solve this problem Einstein had inserted a new term in his gravity equations called the cosmological constant. This antigravity term had allowed the stars and galaxies to spin on happily forever in a fixed, static heaven. The wily Zeldovich seized this cosmological constant and found it could exactly cancel the remainder of the EM energy density remaining from the Planck length cutoff of the ZPF. Ah, what fun to discuss such things on a cold winter's night in Russia over the kitchen table with hot tea, with the aurora shimmering overhead.

Beyond Einstein: GEM Unification

Sakharov discussing politics over his kitchen table.

From these conversations with Zeldovich, Sakharov was emboldened to hypothesize about the foundations of creation. What came before the stars and elements: pure hydrogen. What came before the hydrogen? Pure vacuum. Before the atoms there was the void. The energy of the ZPF was the energy bank upon which the cosmos drew to make hydrogen. Therefore, energy was conserved in a sense. Hydrogen is a plus and a minus charge, an electron and proton which exactly cancel. Therefore, the vacuum could split into a pair of plus and minus charged particles as Dirac had shown. Hydrogen would be that pair of plus and minus.

But Dirac and Heisenberg had shown that the plus and minus charged particles would have to be a particle-antiparticle pair; by quantum mechanics, their masses would have to cancel as well as their charges. Their quantum numbers had to cancel exactly. But that could not be true at the foundation of the cosmos, Sakharov knew. The electron and proton of hydrogen were not antiparticles; one was light, a lepton, and the other heavy, a baryon. They could not cancel. Nature must prefer matter over antimatter at creation, for the cosmos to exist as it did now. Therefore, at its creation, the cosmos preferred matter over its twin, antimatter. Like God nature preferred Jacob over Esau. For if it had been otherwise the cosmos would have annihilated back to vacuum once it was created. But even more sophistication was required to create a universe of hydrogen from nothing.

The protons and electrons popping out of the vacuum meant that not only was matter being preferred to antimatter but that lepton and baryon numbers were canceling numbers at creation. In the moment of creation, the electron and proton had acted as each other's antiparticles, popping out of the vacuum like a particle-antiparticle pair, then somehow not being able to return. Creation had been an irreversible act. A stone cast into the water, making radiating ripples that could never be called back. But what of the ripples of the void, gravity?

Sakharov seized upon his friend Zeldovich's invention of the Planck cut-off and combined it with his work on the Russian hydrogen bomb. The ZPF was like a sea of light around us, and the matter was like the sphere of hydrogen in the casing of a hydrogen bomb. In the blinding instant Sakharov saw that the force of gravity was a Poynting force of variations the ZPF. The ZPF was the energy and the action of Hilbert gravity was the substance of the ZPF.

So it was, that Sakharov, in one period of cold winter's nights, drinking tea with Zeldovich as the aurora shimmered outside, solved the two great riddles of physics, the riddle of hydrogen, and the riddle of gravity, and both came from the ZPF of Einstein. He was very close to the answer of the Great Riddle. Had he pressed on he would probably have found it. However, he published both great ideas separately and never joined them.

Beyond Einstein: GEM Unification

Sakharov became increasingly distracted by the contradictions of the Soviet state he served, and rode off to fight another battle, this one not of physics but of human affairs. He would be one of the leaders of the movement to bring true democracy to the Soviet Union.

Progress was made on another front by Steven Hawking in the 1970s, who connected gravity to entropy. Gravity in the cosmos is the product of large numbers of particles, most of them in hot stars. A star is made of plasma which represents matter in a high state of entropy. Hawking realized that the surface area of a black hole could represent entropy, since it grew in time for all stellar objects. The gravity at such a surface could represent the temperature. He then found something interesting when he shrank a black hole to subatomic size. He found that a quantum plasma formed at the surface of the black hole and allowed particles and antiparticles to tunnel out of its surface. The smaller the black hole, the faster this process occurred. Finally, at the Planck scale, the black hole would disappear in a light transit time of the Planck length into a cloud of evenly matched particles and antiparticles. This matter-antimatter cloud would become pure gamma rays as the particles and antiparticles annihilated each other. Subatomic black holes would thus decay like radioactive nuclei into gamma rays. But what could decay? The matter that had formed the black hole was gone, crushed out of existence. The black hole was pure gravity. What was in fact happening was that the black hole, made of pure spacetime curvature, was transformed into pure EM energy in the form of gamma rays. Hawking had found that gravity could become pure thermal EM radiation. It was the triumph, at least on subatomic scales, of light over darkness.

Beyond Einstein: GEM Unification

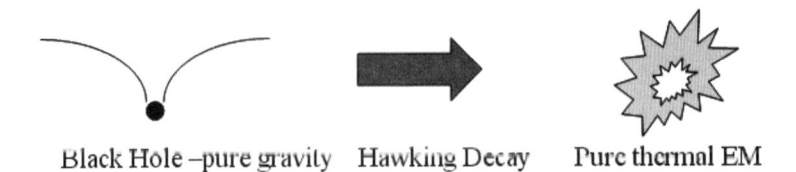

Black Hole –pure gravity   Hawking Decay   Pure thermal EM

The Hawking decay of a microscopic Black Hole, from pure gravity to pure EM radiation.

Steven Hawking.

Later, Gerald t'Hooft, who would share the Nobel prize for physics in 1999 with his professor Martinus Veltmann for studies of the electro-weak theory, studied the decay of particle-like entities called "instantons" into thermal EM fields and would derive an approximate formula for the Newton gravitation constant, G. The formula was exponential, like the normal curve, with the exponent of 137.

Beyond Einstein: GEM Unification

Gerald t'Hooft.

Beyond Einstein: GEM Unification

Chapter 8: Tesla's Vortex - the Cliffs of Zeno

"Anyone who has never made a mistake has never tried anything new."

Albert Einstein

Tesla was born in Croatia, across the Adriatic from Italy. He was born in July of 1856, twenty three years before Albert Einstein. He was Serbian, and Greek Orthodox. He made the world that Einstein grew up in, affecting him from the moments of his birth. Einstein's father had been part of the gaslight-to-electric-light transition was sweeping Europe; however, Hermann Einstein bet on the wrong horse.

Beyond Einstein: GEM Unification

Tesla in his place of power.

Thomas Edison invented the electric light and the concept of electric power distribution, but he had based this system from beginning to end on DC (Direct Current.) This had limited the distance over which electricity could be sent by wires because the copper wires consumed part of the power as heat for every yard the wire was strung. Edison had hired Tesla in 1890 to improve the design of his generators and motors, in return for a large bonus. However, when Tesla accomplished this task, Edison refused to pay him. Edison said that he had been joking when he had promised Tesla a bonus, and that *"Tesla did not understand American humor."* Tesla was too much a genius to be kept down by this petty dishonor; in the end, his triumph over Edison and the systems he had created would be total. Tesla would invent the Poynting vector vortex.

The idea of a vortex of Poynting vectors was entirely the fruit of Tesla's genius. Using alternating current, Tesla created two orthogonal magnetic fields that were in just the right phase so that the magnetic vector rotated in time. A disk of metal, or armature, placed in the rotating magnetic

Beyond Einstein: GEM Unification

field rotated with it because the metal did not want to move across the rotating lines of magnetic force. In a perfect system the disk or cylinder of metal would rotate so that, relative to its own motion, the magnetic field was constant and stationary. Looked at in terms of Poynting vector, the dynamic-crossed E and B vectors formed a vortex in which the metal armature spun. This meant that one could build an electric motor that ran without any electrical contacts between the rotating armature and the magnetic coils around it. The motor ran without the blue light of the plasma that formed the rotating electrical contacts of DC motors and was much more efficient. The new motor was called the induction motor.

The induction motor ran entirely on fields, required no sparking commutators, and could only run on AC (Alternating Current). With this invention, Tesla triggered a "War of the Currents" with the DC empire of Edison, his former, humorous, employer. Allied with Westinghouse and aided by the new efficient electrical transformers designed by Steinmetz, Tesla's vortex and the AC power system it utilized conquered the world. Tesla provided the genius, Westinghouse supplied the financial muscle. Soon, AC transmission towers carried three high voltage wires everywhere, one for each phase of the triangular 3-phase system Tesla invented to optimize his induction motor systems. The 3-phase system required only three wires and a local ground to run and thus saved copper. The secret of the three phase system to create rotation was the principle of an equilateral triangle that formed a plane in which the fields rotated. Soon, the rotating dynamos near Niagara Falls generated the rotating 3-phase AC power and transformed it into high voltage power, which was shipped via three-wire transmission lines to Buffalo, New York to light electric lights and create Poynting vortices in induction motors.

Beyond Einstein: GEM Unification

Edison, after a Herculean struggle, finally conceded defeat, to the man he had once defrauded. One of the losers of this struggle was Einstein's father, who had based his business on Edison's DC system.

3-phase induction motors.

Simplified diagram of Tesla's 3 Phase AC induction motor, where the armature is spun purely by induced currents created by the Poynting vector vortex.

Beyond Einstein: GEM Unification

The world of science in Croatia and Serbia was small, and friendships and personal contacts were very important. It was an especially small world for a remarkable woman who could master mathematics and physics. This woman was a Jewish girl named Mileva Maric. Inspired by Tesla, now a famous son of Serbia, who reportedly befriended her, this remarkable woman journeyed to Switzerland to go to school, where women had a better chance of succeeding in the sciences. Einstein met her and fell in love with her. Even as Tesla had found an ally in Westinghouse now Einstein found in Mileva *"a creature who is my equal and who is as strong and independent as I am."*

Tesla, however, was a staunch believer in the aether, the fifth element and substance of spacetime. Einstein's theories had first made it seem irrelevant and then had made it curve. Tesla was hostile to Einstein's ideas and proposed his own ideas of gravity, based on particles being the subatomic vortex motions of aether. Tesla even said that he had produced what he called "scalar waves" of electromagnetism in the aether. These were like pressure waves in air and, unlike electric magnetic fields in ordinary EM waves, these waves had no vectors associated with them, except for the direction of their motion. That concept of gravity was not a force and due to the geometry of spacetime seemed ludicrous to Tesla and he became a harsh critic of Einstein's gravity theory.

Beyond Einstein: GEM Unification

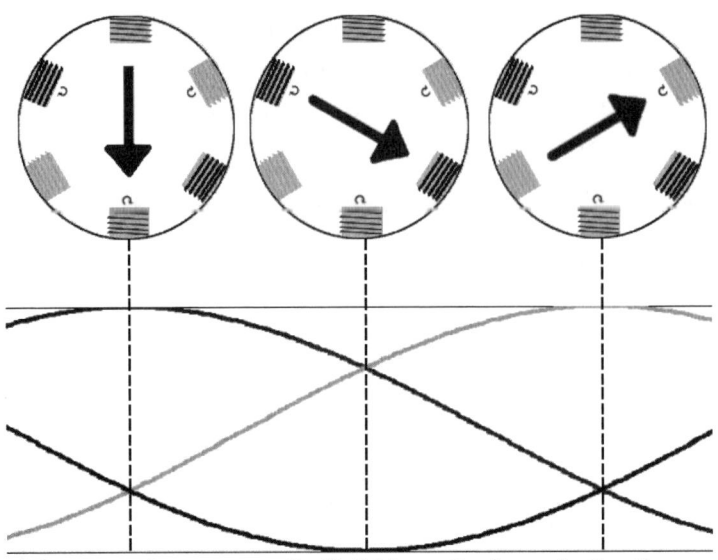

The rotating magnetic field in a 3-phase induction motor.

However, Einstein's star rose as Tesla's star fell. Tesla's last great project, to transmit power without wires, lost its financial backing from J.P. Morgan. Tesla, his mind ever fertile, retired to his hotel suite alone. He proposed that gravity could be controlled by electromagnetism in 1938, but his personal eccentricities had by this time undermined his credibility. Einstein followed Tesla's genius and invented a linear induction pump for a refrigerator called the Einstein-Szilard pump. It is also rumored that Einstein and Tesla collaborated in World War II at the Philadelphia shipyard on experiments to render warships invisible to radar, but if this occurred, the experiments and their outcome remained secret.

Einstein pressed on, however, in his final great quest to unify electromagnetism and gravity. His program involved not only to unify the fields but to explain the existence of hydrogen: the proton and electron. Thanks to Einstein, gravity fields were now fully dynamic and supported waves

Beyond Einstein: GEM Unification

like EM fields. To Einstein, as to Tesla, the similar mathematics and structures that gravity and EM created in the world argued they had to be fundamentally related. Both gravity and EM followed the inverse square law, both made waves moving at the speed of light. The mathematical connection went even deeper, now that Einstein had made time a fourth dimension.

High voltage power transmission. Note the 3 lines mounted on glass insulators that carry Tesla's 3-phase power.

Force fields now existed in four-dimensional space. They were tesseracts, things that could no longer be visualized directly, only seen as three-dimensional shadows. To see the variation of a field of temperature or pressure in three-dimensional space one would see a field of numbers. Temperature and pressure are physical quantities that are represented as a single number in space. Such physical quantities are called scalars. To see the variation of the scalar field was to look for the direction of its greatest rate of increase. This is called the gradient vector. To experience a gradient one need only walk straight up hill or watch water flow down one. Therefore, the pattern of variation of a scalar field in space is a vector, which is a set of numbers describing rates of change in space, but the pattern develops further.

Tesla's AC dynamo and motor on display in 1893.

To describe the variation of a vector field most fully, such as in the distortion of a solid under stress or a water flow, we use a tensor. Tensors were invented to describe the pattern of distortion of solids under pressure or tension. They are written like matrices with rows and columns, whereas vectors were written as merely a row of numbers and a scalar was merely a single number. However, a vector is easy to imagine in three dimensions as an arrow. A tensor is more difficult. The Einstein equations of gravity were all in tensor form, expressing the tidal distortion of matter in a gravity field and by this the curvature of space. The basic mathematical form of the theory was that of the "metric tensor" which expressed the structure of spacetime. The equations of Einstein were written in terms of this metric tensor and its curvature, which also was a tensor. Unlike

Beyond Einstein: GEM Unification

Maxwell's equations, with easy to visualize vector fields, Einstein's equations seem much more abstract. However, some tensor fields can be visualized simply.

The simplest tensor expressing a variation of a vector field is that representing a vortex flow. It is a matrix with diagonal made of all zeros and an upper half which is the negative of the lower half. It is easily recognized and easy to visualize. It to this concept of the EM vortex that Einstein, an intuitive thinker, appears to have been drawn to trying to unify the fields.

Minkowski, a fellow student of Einstein's, introduced the idea of four-dimensional spacetime; this led quickly to writing the EM field in four dimensions. The electromagnetic field is no longer a vector in four space; it is a tensor, and this "Faraday tensor" describes a vortex in spacetime.

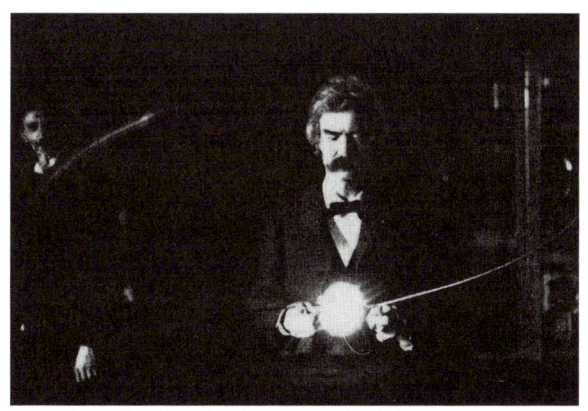

Mark Twain in Tesla's laboratory.

The vortex appears everywhere in physics because nature loves the vortex. Everything spins in nature: the subatomic particles, the electron around the proton in the atom, Earth spins and orbits the Sun, the galaxy is a plasma of stars that spins. The rotation of a solid body is rigid, with the same

Beyond Einstein: GEM Unification

rotation frequency throughout. But most matter in the universe is a fluid plasma. Such flows form a vortex where the rotation frequency varies throughout the plasma. With the structure of spacetime and its spacetime as a tensor and the EM field as tensor Einstein saw the way clear to unify the fields. So he was mesmerized by the vortex tensor of EM.

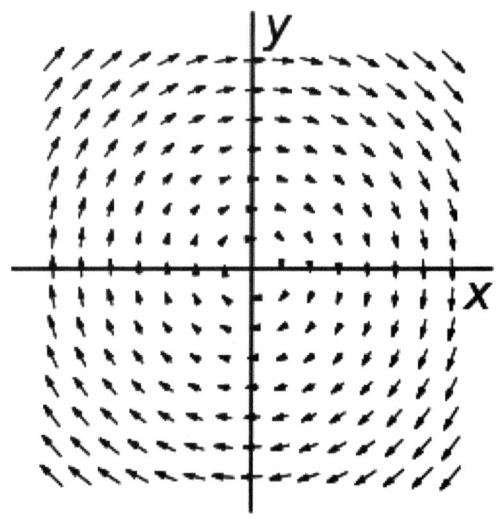

A swirl or vortex pattern of vectors.

The fields of gravity as a spacetime tensor and its curvature, and the vortex tensor of EM must be related mathematically, thought Einstein. The effect of gravity alone was to produce a tensor that expressed a stretch and squeeze in spacetime. However, the effect of an EM field was to create a swirl in spacetime. This seemed obvious to Einstein. Did not the charged particles of a plasma move in a vortex around magnetic lines of force? The vortex field of EM must be a vortex like Tesla's vortex moving everything within it. The paths of spacetime must be twisted into a vortex in the presence of an EM field.

Beyond Einstein: GEM Unification

The spin of subatomic particles had been described by Pauli. He found the spin was expressed by vortex matrices also. To Einstein it seemed all the world of physics, even the world of quantum mechanics that he despised, was waking up to the importance of vortex flow. However, the gulf was wide between Einstein's physical picture and a mathematically consistent theory.

Quantum mechanics requires navigating a sea of paradoxes, such as wave-particle duality, to calculate measurable effects. One can only do this if one avoids being hypnotized by the paradoxes. For this reason, quantum mechanics, more than other branches of science, requires that certain questions be discouraged. However, Einstein did not like being told that certain questions were nonsense. His naturally rebellious nature made him keep asking questions. He stared the paradoxes of quantum in the face and continued to demand that someone explain them. One problem he continued to wrestle with was what held the electron together.

To Einstein, a true unified field theory would explain the riddle of the electron and other particles. He became convinced that gravity must hold the electron together. EM wanted to explode the electron, gravity wanted to collapse it to a black hole, and somewhere there was an equilibrium. But how could this be?

To have a finite mass yet be a point it was obvious that the electric field at the electron surface quit getting stronger at some radius as it shrank. This is called saturation in EM, and occurs in things like magnetism. In iron, for instance, the magnetic field of a bar magnet will reach a certain intensity and before it can contain no more magnetic flux lines, like a sponge so full of water it cannot contain any more moisture. So, the electron is a sphere of charge compressed to smaller and smaller radius, the electric field

Beyond Einstein: GEM Unification

would become more intense, and compression would get more difficult. But suppose the electric field saturated at some radius, as if the vacuum could carry no more electric field lines, Then the compression would not become more difficult even if the electron did shrink to a point. The mass would keep growing and also reach a maximum. So field saturation could reach finite mass for the point electron. Quantum mechanics, unfortunately, does not predict any saturation, only that the electric field will not increase as strongly as the electron shrinks.

What of gravity however? The gravity at the surface of the electron need not saturate, but could grow continuously as the electron shrinks. Finally, gravity would become very strong as the electron's radius dropped to zero, even as the electric field and mass saturated. Finally, at some tiny radius, assuming electric field saturation to make the electron mass finite, the gravity would be as strong as the electric repulsion and the two forces would balance. The electron would be held together by gravity at some tiny radius. With some spin on the electron surface, the electron could not collapse into a black hole, but would be stable. It would have to spin to exist. So, if one found a saturation of the electric field at the electron's surface, then the existence of the electron, and its inherent spin, could be explained.

But though Einstein looked high and low for such a saturation effect, nothing in either Maxwell's equations or quantum mechanics suggested that a saturation of the electric field at the electron surface was possible. He tried to find this effect in his unified field theories but instead found himself ensnarled in complex math that even he could not unravel.

The tensors of gravity were the metric tensor and the curvature tensor, and they were much more complex than the simple EM vortex. The gravity field so familiar to anyone on

Beyond Einstein: GEM Unification

Earth had disappeared in the tensor theory of spacetime. Gravity was not a force in Einstein's theory, but was an illusion induced by the curvature of spacetime. In a closed elevator one could gravity of acceleration was a work. In a freely falling elevator one becomes weightless and gravity disappears, at least until you hit the bottom of the shaft. The only way to tell gravity from acceleration is to look at patterns of small masses as they drop. If the pattern stretched in the direction of its fall and squeezed in the plane across it, gravity is at work. If the geometrical pattern was preserved, then what looked like gravity was merely an acceleration incurred by the cable of the elevator being pulled. Only curvature was real in Einstein's theory, but curvature was not a vortex but a stretch and squeeze pattern, or in some cases, a simple shrink in all directions. When water drains in a sink it forms no vortex unless it spins first. If the water is still when the stopper is pulled it drains without a vortex forming. The vortex we see on a large scale, the maelstrom, the tornado, the hurricane is due to the rotation of the Earth. Get rid of the rotation, and gravity still works just fine.

Beyond Einstein: GEM Unification

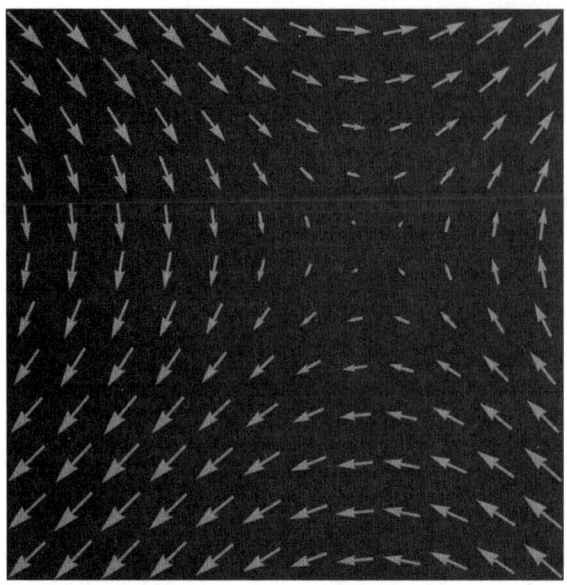

The "stretch and squeeze" or tidal pattern of vectors characteristic of gravity.

EM forces are real, they move things. They will not go away in an elevator, they are made by electric charges and their currents. A magnet still attracts iron whether or not the elevator breaks its cable. So, for Einstein to unite the gravity field with the EM field, he had to unite a force with an illusion. Vortex or squeeze-and-stretch, it was a difficult and slippery hill for Einstein to climb. Finally, it would become a sheer cliff that was smooth as glass. Einstein labored on despite the difficulty and his increasing isolation in the physics community. He had to unite the Faraday vortex tensor with the stretch-and-squeeze tensor of spacetime curvature, but not even their units matched. How could one add eggs to potatoes?

The Cosmos, with galaxies like dust. (NASA)

Stellar explosion remnant with expanding plasma on magnetic filaments. (NASA)

A rainbow showing the continuous spread of the visible EM spectrum of the Sun. (NASA)

The line spectrum of thin hydrogen plasma.

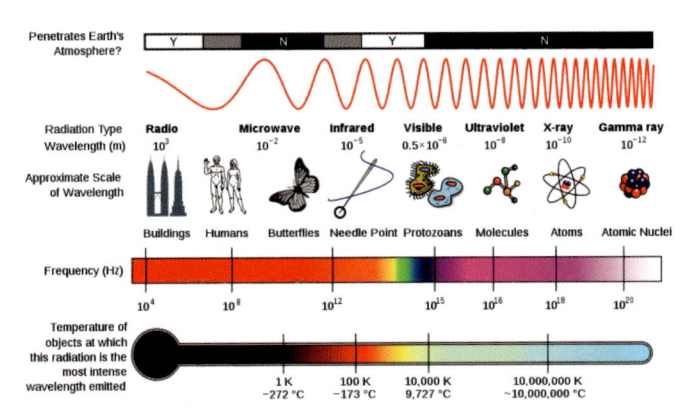

The Electro-Magnetic spectrum, of which only a fraction is visible.

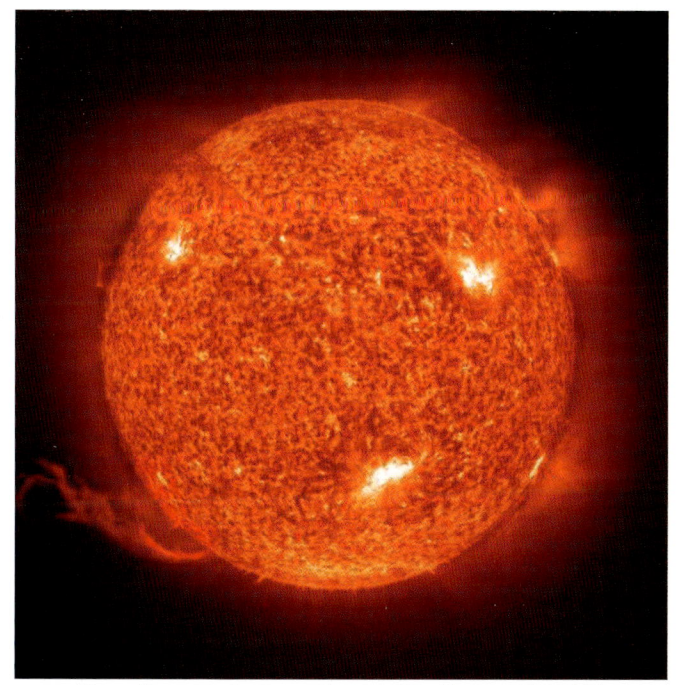

Gravity creates a sphere: the Sun. (NASA)

EM quantum forces create a sphere: a soap bubble.

Gravity creates a vortex: a Spiral Galaxy. (NASA)

EM and aerodynamic forces create a vortex: a Hurricane. (NASA)

Magnetism: Plasma clings to arched magnetic lines of force on the Sun.

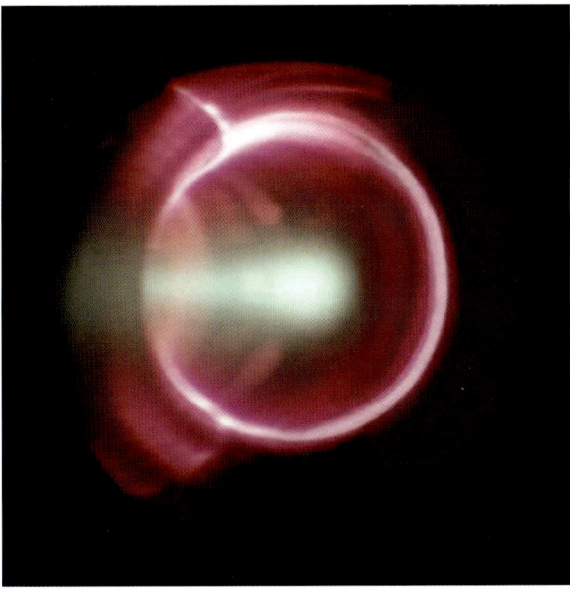

Magnetism: Plasma clings to lines of force in a Tokomak, invented by Andrei Sakharov. (DOE)

Electricity: plasma follows lines of electric force from a Tesla coil.

Electricity: Lightning- Plasma forms on electric lines of force from the sky to the ground. (NOAA)

The Pillars of creation: interstellar dust is sculpted by EM radiation pressure. (NASA)

Gravity and the Strong Force: the stars. (NASA)

Magnetism: the Aurora viewed from Earth's surface.

The Aurora viewed from Space, electricity flowing down magnetic lines of force. (NASA)

Beyond Einstein: GEM Unification

Tesla's wireless power transmission project.

In units the bare mathematics of theoretical physics meets the harsh reality of real life experiments. One can contrive, by some physical model potatoes harvested in a field chopped up, dried, and fed to chickens, to equate potatoes grown to eggs laid, but somewhere in the model must emerge a conversion factor of units of eggs to potatoes. Both potatoes and eggs can be counted, but they must be categorized. Many physicists of a mathematical bent are pleased to convert all their equations to "unitless" ones; that is, into pure numbers. On this path lies great danger, for the physics must match reality, and reality runs on units.

Beyond Einstein: GEM Unification

Einstein, Tesla and Steinmetz meet in 1921.

For Einstein the goal seemed breathtakingly close. The tensor of the curvature of spacetime has the units of inverse square of distance, as does the force of gravity itself. The Faraday tensor also has the units of an inverse square of

Beyond Einstein: GEM Unification

distance, but then this is multiplied by the units of electric charge. Einstein had to unify the units of EM and gravity to have any hope of unifying the fields. The conversion factor to make spacetime curvature into EM field and vice-versa had to be a factor that eliminated the charge from the Faraday tensor units. The units had to be an inverse charge unit. But simply dividing by the charge of the electron yielded nonsense. Some factor in nature had to exist that was characteristic of the universe that had natural units of inverse charge. By taking a charge, dividing it by a mass and the speed of light squared, and then multiplying by a length, he could achieve the right units.

What Einstein was attempting to do over decades boils down to the search for a universal physical constant that could give him this factor of inverse charge. Nature seems to eagerly offer part of the answer for this factor's components. The choice for the universal charge was obviously the electron charge. However, a universal mass was more confusing, it must be mass of the proton or the electron, but which? One can proceed pragmatically by reasoning, like Einstein, that since there are only two obvious masses, those of the proton and electron, it can be sorted out by seeing which mass works the best, since they are only a factor of 1836 apart. The speed of light is universal, but when time and space are unified in relativity, they now have the same units, so the speed of light in four-dimensional physics is unit-less. What remains is a universal length. Since Einstein's theory of gravity had been geometric, it seemed only reasonable that the unification of gravity and EM would boil down to some universal length. However, the length must be in vacuum itself, for if the atoms provide the charge and mass, the void must provide the final key piece of information for unification. What length was hidden in the void?

Beyond Einstein: GEM Unification

Einstein later in life, still trying to unify the fields.

Einstein needed a bridge across the void that separated EM and gravity, and the length of that bridge was the fundamental length that would finish his theory. But in the classical vacuum he contemplated was only emptiness. He rejected quantum mechanics and its vacuum, full of virtual particles, with the same stubborn independence that had led him to propose the theory of relativity decades earlier. Only atoms and void existed and the atoms had size to them and structure, but in the void was nothing. By the paradox of Zeno, the distance between Achilles and the tortoise could be infinitely subdivided. The classical void was smooth as glass. It yielded no length for Einstein to get a foothold on. So it was that Einstein fell from the smooth cliffs of Zeno as he tried to climb them. Down he fell into the Vortex of Tesla, and shared his fate, dismissed and forgotten by the

Beyond Einstein: GEM Unification

physics community, his quest of achieving the unification of gravity and EM dismissed with him.

Tesla studying at one of his coils.

Beyond Einstein: GEM Unification

Chapter 9. The Hidden 5th Dimension

It is reported that Gunnar Nordstrom, a promising young physicist from Finland, was the first to realize that Maxwell's equations came directly from the presence of a 5th dimension. He was also the first to link gravity to the metric of spacetime. Tragically, however, he died at an early age, apparently from ingestion of radium, which was considered a health aid in those days.

Gunnar Nordstrom, the father of the 5th dimension.

Beyond Einstein: GEM Unification

Theodor Kaluza.

Theodore Kaluza was born in the German city of Kronigberg, now named Kalingrad. He was a talented mathematician, but rather shy and retiring. Like Einstein, he was attempting to unify gravity and EM. In 1919 Kaluza made a remarkable discovery concerning the Hilbert action principle. The Hilbert action principle basically minimized energy in spacetime as if it was a block of rubber under stress. The resulting relaxation equation distributed the stress throughout spacetime in the way that led to minimum total stress. David Hilbert had discovered that when one did this in the four dimensions of spacetime then Einstein's field equations for gravity would result. However, as Kaluza explored further, he made the spacetime five dimensional, with the new dimension being something that did not appear in normal situations. It was there but it was not. The new 5th dimension was hidden. What resulted from this mathematical procedure was astonishing.

When one minimized spacetime curvature stress in four dimensions, one obtained the equations for gravity, but when one did this in five dimensions Maxwell's equations for

Beyond Einstein: GEM Unification

electromagnetism also fell out together with the Einstein equations for gravity. Nor did the equations fall out as separate systems, for the source term for the gravity equations was the energy density tensor of electromagnetism, called the Maxwell stress tensor. It was an amazing mathematical result that was all the more astonishing because the mathematics was so simple. The boundary between geometry and forces was now gone, EM was geometry in five dimensions, and gravity was a force. The fields could now be unified. The theory also predicted a scalar type of wave, half way between electromagnetism and gravity.

The implications of the Kaluza theory seemed enormous. The equation of special relativity equating mass to energy was seen as ingrained in the structure of spacetime itself, with, in this case, the energy density of the electromagnetic field serving as the mass density that would create gravity. Kaluza, like many researchers who found new and radical results, sent the paper to Einstein, who had become the unofficial expediter of publication for new and strange ideas. As troublesome as this sometimes seemed, this role for Einstein hearkened back to his role as a patent examiner in Switzerland, and also allowed an outlet for his rebellious tendencies. If he could not stir up the pot of physics himself, he could help others do it.

Einstein initially viewed Kaluza's paper with suspicion, however. It seemed too good to be true. Einstein's own work on general relativity had only recently been supported by solar eclipse results. Einstein had based his theory on a strong sense of what was physical, but the Kaluza result seemed very abstract and motivated by mathematics rather than physics. It made no sense to Einstein that a hidden 5th dimension could exist, have such profound consequences, yet not be detectable. He had similar problems with the concepts of quantum mechanics. Einstein's mind rebelled at the idea of something hidden yet so important. However, he realized the Maxwell equations were well verified, better verified than his own equations. That the same mathematical procedure yielded both sets of equations and their connecting term, the Maxwell stress tensor, gave validity to his own work. Einstein sat on the article for nearly a year, but, perhaps emboldened by the increased support for his gravity theory, he then decided to endorse it for publication. So in 1920 Einstein made possible the publication of a theory that unified gravity and electromagnetism mathematically, yet a theory he himself rejected as unphysical and spurious. Einstein continued on his own approach to the problem of unification of gravity and electromagnetism. The physics community as a whole seemed even more suspicious than Einstein of this new idea. However, the theory of Kaluza did not go unnoticed by one physicist, who decided to develop it further.

Beyond Einstein: GEM Unification

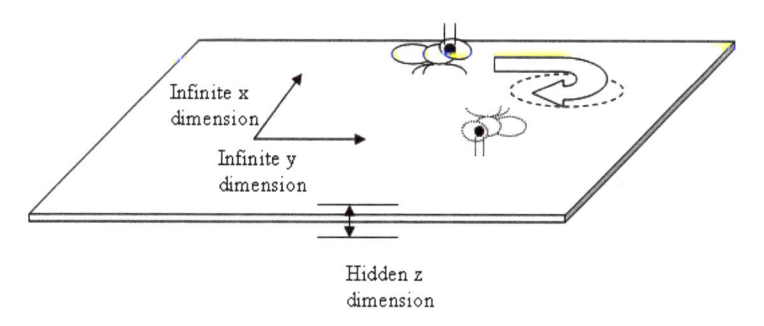

An ant experiences a hidden dimension on an endless sheet of paper. The ant can move in the x and y directions to infinity, but can move only + or - half the width of the sheet of paper in the z direction. Alternately, if the ant can move though a hole in the sheet, it will experience a negative z displacement of half the width of the paper.

Oscar Klein, the son of a Swedish rabbi, studied the theory of Kaluza while sick with hepatitis and was able to reformulate it. He ignored the scalar wave portion and set it to be a simple constant. What was newest about Klein's approach was the interpretation of the hidden 5th dimension. Klein was not content to have it simply be a place holder like Kaluza but had it "curled up" in a circle with a small radius. The 5th dimension would thus have an inherent length. The new formulation gave the equations of quantum mechanics and showed that the radius of the new curled-up dimension related Planck's constant to the charge of the electron. His work in polishing and focusing the theory made it much more physical; hence, the theory has been called Kaluza-Klein ever since. However, the theory did not predict any new phenomenon, it simply explained what was already seen in a bizarre way. Einstein went down his own path to unification without embracing it. The theory languished as merely a mathematical oddity. But the memory of the physics community is long.

Oskar Klein.

As physics progressed in the United States the main effort became focused on the strong force and the nucleus that it held together. The strong force determined the properties of the nuclear particles, the nucleons: neutrons and protons. Great advances had been made.

The electron had been studied carefully. It seemed to have no real size but instead a point-like form; however, the nucleons, the proton and neutron, seemed to have a size and inner structure. The proton and neutron had been found to form a liquid-like entity in the nucleus that was incompressible. This meant that each proton and neutron occupied a definite volume. Since the simplest shape for the proton and electron seemed to be a sphere; this meant the proton and neutron were of the same radius. Despite the fact that everyone considered the electron to a perfect point, the nucleon radius, by coincidence, appeared to be almost exactly the radius given to the electron based on how it scattered light. It was the radius an electron would have if its mass was due to pure electrostatic energy caused by its charge spread uniformly in a thin layer on an otherwise massless shell. No theory would connect the two things, the radius of the proton and the hooked-up radius of the electron that was really a point. The radius of the proton was real and could be measured. This radius seemed to be the length over which the nuclear glue, the strong force, acted.

The strong and weak forces of nature, unlike the forces of gravity and electromagnetism, are of limited range. They do not obey an inverse square law but instead act strongly over short distances and then become zero. This behavior reminded someone of the Kaluza–Klein curled-up dimension.

Strangely, Wolfgang Pauli, usually the most skeptical of physicists, took a liking to Kaluza-Klein theory and kept its memory alive. In 1953, the year the author was born, he corresponded with Abraham Pais concerning its use to describe the strong force between nucleons. This was a natural impulse. Kaluza-Klein theory was not only a natural thing to try to describe the strong force because it introduced new coordinates independent of x, y, z, and time, it also introduced an embedded length that was the size of the hidden dimension. The strong nuclear force has a natural scale size. Whether by design or coincidence, the scale size of the nuclear force is the classical radius of the electron. If one considered this the size of a hidden dimension, where it could mix freely with x, y, and z, then a Kaluza-Klein theory seemed a good place to start. The length scale of the strong force interacts with the quantum vacuum in a way that the classical vacuum cannot. As alligators are attracted to

Beyond Einstein: GEM Unification

isolated bodies of water to call their own, any preferred length in Kaluza-Klein must attract a particle to inhabit it. The mass of the particle will be determined by the hidden dimension size, so its wave nature is in resonance. In the case of the strong nuclear force with its characteristic length, the particle is the π meson or pion. The existence of this particle was predicted by Hideo Yukawa in Japan. The meson comes in three varieties of electron charge, plus, minus, and neutral. They last only a billionth of a second and decay into electrons and pure light. They dash in between the nucleons to keep them together even while their positive charges try to force them apart.

Following Pauli's lead, various attempts were made to make the strong force fit into a Kaluza-Klein framework through the 1960s, but another more powerful theory about the strong force had emerged.

Science had become the right hand of the government in the 1950s and 60s. The probing of the atom had yielded undreamed of power to the nation's arsenal in the form of nuclear weapons and the Cold War was raging, so the probing ever deeper into the subatomic realm continued on both sides of the Iron Curtain. As the proton and neutron themselves were probed with collisions of higher and higher energy in particle accelerators, a "zoo" of strange, short-lived particles flew out like streamers from a large fireworks and would explode themselves after a few billionths of a second. The deeper one probed, the more complicated things appeared. But order was seen in the apparent chaos.

Murray Gell-Mann and George Zweig of America, and, independently, Yuval Ne'eman of Israel arrived at the solution of the puzzle: The proton and its cousin the neutron were triune. They contained each three sub-particles yet were indivisible. The three particles were called quarks, a word rhyming with sparks, but whose origins arose from the

Beyond Einstein: GEM Unification

epic of Irish literature *Ulysses* by James Joyce. This triune structure meant that the subatomic particles could become excited internally like atoms, but unlike atoms they could not break up into a plasma; the quarks were bound together inseparably. This epic achievement of theory was confirmed by the new particle it predicted, aptly named the Omega-minus. Taken together the family of particles looked like a large crystal, with subatomic particles at its vertices.

The strong force fields held the quarks together, and the fields were called color fields, with the quarks being red, green, and blue. They are believed to be like EM fields, but they are strange in that they swap color charge between the quarks as they create dynamics between them. As the electrons radiated white light, the quarks formed a particle of colorless light that scintillated like a star as the quarks changed colors.

It was discovered in the 1970s that particles could be modeled as vibrating strings. The particles held together by the strong force, such as the proton or neutron, behaved as if they were made of internal subunits, and these subunits would engage in quantum levels of excitation like a vibrating string. Increase the tension on a string and the note it vibrates at will increase in frequency. Some of the theories used closed loops, some used open string segments. String theory was thus born. It also became apparent that the theories worked best when more than four dimensions could be used, as if the nucleon was a doorway to a more complex universe. The proton is different from an atom, which comes apart to form a plasma when it becomes too excited; the proton on the other hand would not come apart but instead behaves as if it was accessing new dimensions. Then another discovery was made; a string solution that appeared in the theories was the graviton.

Beyond Einstein: GEM Unification

The graviton is a particle similar to a photon, a particle of EM energy, but instead of an EM field, the graviton carried gravity. These particles were first isolated mathematically by Feynman, when he attempted unsuccessfully to fully quantize gravity. Feynman found that quantized gravity only worked for nearly flat spacetime, that is, weak gravity fields. When strong gravity fields were subjected to his mathematical procedures, they were successfully used for EM fields; the equations of gravity created black holes and infinities. However, in the world of subatomic physics, spacetime was always nearly flat, so the weak gravity quantum theory would work. This meant the string theory of particles was now also a theory of the spacetime around them.

The person who saw this most clearly was Edward Witten, who wrote extensively on the application of Kaluza-Klein theory to the string problem. The theory of strings had now become the theory of many curled-up dimensions. The theories varied, requiring many more dimensions than the original five proposed by Kaluza and Klein. The goal of all of them was the same, to unify gravity, EM, the weak, and the strong forces in a theory that explained both space and particles. String theory had become a theory of the atoms and the void, a Theory Of Everything: "the Big TOE." However, there is a famous saying of Sun Tzu, from *The Art of War,* "To try to defend everything is to defend nothing." So the theories of everything fell short.

In the middle of this tumult, other unified field theories had emerged based on unification near the Planck energy. These theories attempted to unify the strong, weak, and EM fields and were called Grand Unification Theories or GUTs. Some of these theories made the remarkable prediction that the proton would decay like the neutron, leaving only a positron to carry its positive charge. This decay would take a

Beyond Einstein: GEM Unification

long time, so the stage was set for the most boring experiment in history.

A mile underground, shielded from all cosmic rays that might interfere, a vast chamber was filled with a million tons of ultra pure water. Thousands of sensitive light detectors were pointed into the clear vessel to watch its crystal clear water in utter darkness. If the proton decayed in time, then one of the countless billions of protons in the water would pop like popcorn, releasing charged particles that would create a spark of light. The doors were closed, the light turned out, and the vigil begun. Then remarkably, nothing happened, even as years passed.

Finally, one night in 1987, the tank lit up with numerous sparks of light. It was as if fireworks had gone off. The researchers were initially delighted; it appeared for a moment that a proton had gone off like a bomb. But their elation was short lived. Examination of the data and reports from similar experiments showed that the phenomenon they had observed was not from deep within the proton but from the deep outer cosmos. A supernova had gone off in the nearby dwarf galaxy called the large Magellanic Cloud and had showered Earth with neutrinos in such numbers that several of them had collided with protons and broken them apart in the tank. It was as if Mother nature had turned prankster, after spying a party of bird watchers listening for the call of some rare species in a park at night, she then had tossed a cherry bomb into the parking lot behind them.

This was strangely reminiscent of one peculiar result of the simple Kaluza-Klein theory, uncovered by Edward Witten. It was quite startling. It seemed The Kaluza-Klein hidden dimension messed up the pristine order of four-dimensional spacetime and made it unstable to quantum processes. The particles were stable, but spacetime was now unstable to decay like a nucleus. Except the result was bizarre; spacetime would decay into expanding mirrored spheres filled with nothingness. So it was that the proton was seen to be stable as rock, much more stable, in fact, but the fabric of spacetime could unravel at any moment. This effect found by Witten was called vacuum decay.

Chapter 10. The GEM Unification Theory

"By knowing things that exist, you can know that which does not exist."
Miyamoto Musashi

"God is subtle, but He is not malicious."
Albert Einstein

We now stand in a flat expanding cosmos whose basic building block is hydrogen and whose structure depends on the ZPF. That much is mainstream physics. But now, having found a purely EM effect that looks like gravity, that involves the all important Poynting vector, and also seeing that a five-dimensional spacetime produces both EM and gravity equations, we can unify the fields. Einstein was right in his basic premise that this unification must involve the existence of the electron and proton. The unified field theory that results will be very profitable to us, both in intellectual and utilitarian ways.

Basically, the GEM theory is this: EM and gravity both operate on the Poynting vector. The basic structure of spacetime is a microstructure of powerful electromagnetism. The ZPF is largely hidden because of spacetime. The ZPF is slightly weakened in the presence of matter and this local weakening creates variations in the ZPF. Variations in the ZPF create "curved" spacetime and thus in gravity, or alternatively, variations in the Poynting force that push matter towards other matter. When the 5th dimension appears it captures some of this electromagnetism at the Planck scale and then deposits it in the subatomic scale in the form of two charged particles, the proton and electron, and this allows the appearance of EM fields separate from gravity fields and also from matter itself.

The GEM (Gravity Electro-Magnetism) unification theory will allow us to create antigravity for flights to the stars, for it will allow us to directly modify gravity fields with EM fields. Gravity, according to Sakharov, is in fact the variation of the Poynting vector field or radiation pressure of the ZPF. The present gravity theory says we must create mass with EM fields to change the curvature of space. That requires more EM energy than is possible technologically, either now or in the distant future. However, we will see that the GEM theory allows us to curve spacetime directly by using its underlying electromagnetic nature. In the GEM theory, we see that spacetime is electromagnetism or more precisely, it is the EM ZPF. This fact also helps to understand the cosmos we journey across.

The GEM theory acronym also stands for the Latin phrase "Grandis Et Medianis," the unity of the great and middle scales of the universe. The GEM theory unifies the great energy scales of the very small, the Planck scale, and the Cosmic scale, with the middle scale of atoms and subatomic particles that underlie reality. We call this middle scale the "mesoscale"; its basic unit is the effective size of the fifth, and hidden, dimension. The GEM theory will tell us why we live in a universe that is essentially flat and full of hydrogen with two long-range forces, EM and gravity. The strong and weak forces of the cosmos are details we will consider later.

Beyond Einstein: GEM Unification

What the GEM theory gives us is a thumbnail sketch of the cosmos as it is. Let us then assemble the pieces of the theory we have studied previously.

We know, because we are now familiar with basic plasma physics, which runs the universe, that the ExB drift affects all particles regardless of their mass or charge. We also know that all subatomic particles that make up matter, the electrons and quarks, are charged, and at a basic level move freely. The ExB drift, consisting of a crossed magnetic and electric field, causes all free charged particles to move at right angles to the E and B field with a uniform velocity. The velocity is well known to all plasma physicists and is called the ExB drift velocity. It also could be called the Poynting drift, because the ExB vector is the Poynting vector, however, we will adhere to conventional plasma physics terminology here.

Imagine two infinite parallel plates at different voltages with an electric field between them. If we add a magnetic field parallel to the faces of the plates then any charged particle will begin to move at the same velocity. The stronger the magnetic field that underlies the ExB drift, the faster the particles will all assume this motion. Now that we have this device for making all charged particles move the same, we induce a variation in this device. We change the voltage difference on the plates with time. The particles all smoothly pick up speed or slow down and all the velocities change in time identically, regardless of charge or mass.

We can do another thing to vary the fields, this time in space instead of time. We simply tilt the plates relative to each other so the electric field gets stronger in the direction of the particle motion, where the plates are closer together. The particles again smoothly accelerate toward the strongest part of the electric field, regardless of charge and regardless of mass. What we have done, then, is make a gravity field

Beyond Einstein: GEM Unification

with electricity and magnetism. It is as simple as that. Gravity can be made with pure EM.

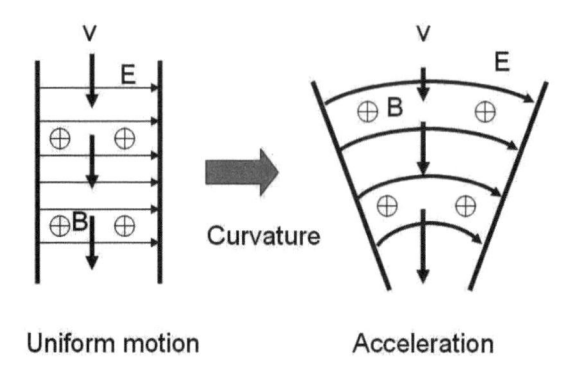

The ExB or Poynting drift of a plasma in crossed electric and magnetic fields.

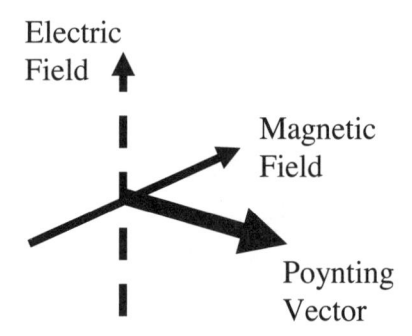

The Poynting vector of electromagnetism.

Beyond Einstein: GEM Unification

Looked at in terms of spacetime, our initial arrangement of parallel plates, fixed voltages and particles is a flat spacetime. When we vary the electric field with time or space to produce the acceleration of the particles, we have curved the spacetime. Thus, gravity is due to the curvature, or variation, of the electric field. The magnetic field, however, is everywhere constant. This constancy of the magnetic field in space and time is important and will be discussed again.

This is how we can make gravity for charged particles in a vacuum chamber in our lab, but how does nature make gravity that we live in? Where are the plates, voltages, and fields that move the particles of our beings? To answer this, we must consider that what we have made in the laboratory is the magnified version of the action of gravity on the subatomic scale. We must consider the nature of particles and spacetime itself.

The classical vacuum is barren and sterile, it is true nothingness. It is like a dreamless sleep.

The classical vacuum, emptiness.

Beyond Einstein: GEM Unification

But such a thing as the classical vacuum does not exist. When we sleep, we dream, and dreams are an important part of our life. Similarly, the quantum vacuum is alive with possibilities. It is the quantum vacuum that the fields are unified.

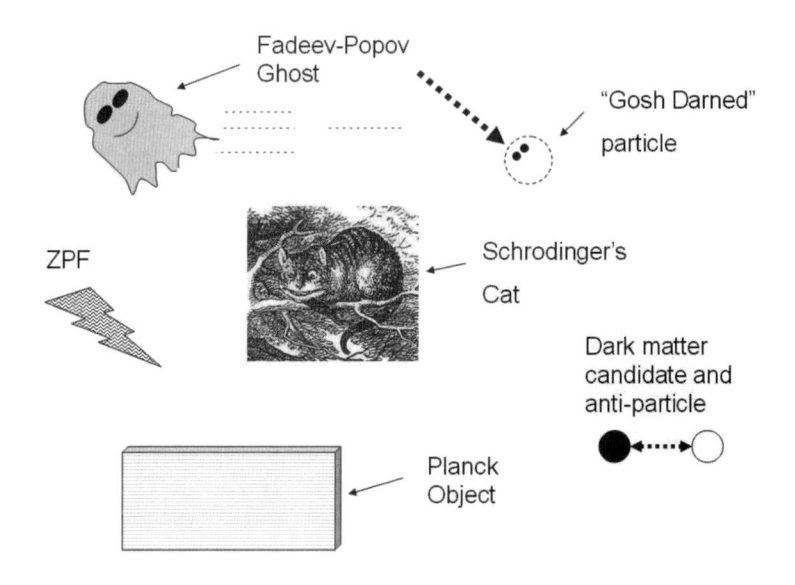

Quantum vacuum: a sleep full of dreams.

So the mechanism of GEM gravity exists in the quantum vacuum, in the ZPF. The ZPF is where the magnetic fields exist, and the charges of the fundamental particles supply the electric fields.

The simplest way to imagine the ZPF is as a white light as from a furnace, but so powerful it censors itself. If it did not censor itself, we could not exist because all matter in the universe would flash into plasma. Self-Censorship occurs because the ultra-strong EM fields of the ZPF *become spacetime* itself. The all powerful thing becomes invisible because it becomes the fabric of reality - it becomes the ground of being, just as a leaf blown in a strong wind feels no wind or a deep sea fish has no sensation of pressure. Self censorship of the ZPF allows us to exist and also ensures that the mass density of the vacuum is zero. The ZPF censors itself by becoming the fabric of spacetime. So that we may exist in a cosmos with a ZPF, the ZPF must censor itself in terms of the mass-energy of the vacuum. Yet the ZPF can be felt in subtle ways: The ZPF controls the decay of excited atoms, it makes the charged particles experience "jitter" according to the Heisenberg uncertainty principle, and if two metal plates are held near each other the Poynting force imbalance between the inner face of the plates and the outer faces causes them to cling together. It is the hypothesis of the author that the ZPF causes repulsion between surfaces of water to produce a remarkable effect called the anti-bubble. If so, then anyone can see the ZPF at work in their kitchen sink.

Beyond Einstein: GEM Unification

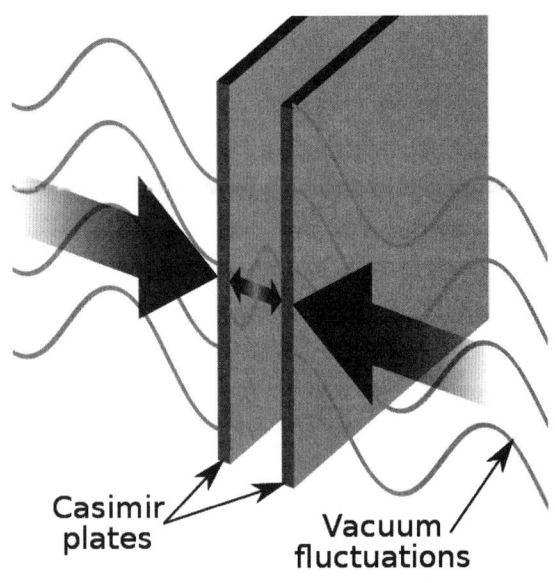

The Casimir effect: Two plates exclude ZPF waves, leading to attraction of the plates because of the increased pressure on the outsides of the plates. The ZPF is thus readily observable in its effects.

In the anti-bubble the parallel surfaces of two strong dielectrics can be held apart by trapped ZPF fields with separations of a wavelength of light. This effect makes the well-known effect of skittering droplets of water moving over the surface of water seen while rowing a boat on a calm lake on a sunny day with one's sweetheart. The bright droplets capturing sunlight on the lake can actually be pushed beneath the water surface and there form an anti-bubble seemingly in defiance of the laws of physics. The anti-bubble consists of a thin layer of air between a sphere of water and the surrounding water and is a marvel to behold. Some powerful force obviously keeps the surfaces apart, and the anti-bubble can last for many minutes in this state. The ZPF is in action in this case, the author has hypothesized, and the effect is very strong due to the peculiar properties of water, its high dielectric constant, and its remarkable window of transparency to EM waves in the visible wavelengths, while being black as ink in the infrared and ultraviolet. Therefore, if the author is correct, then the power of the ZPF that creates gravity can be seen in a glass of water.

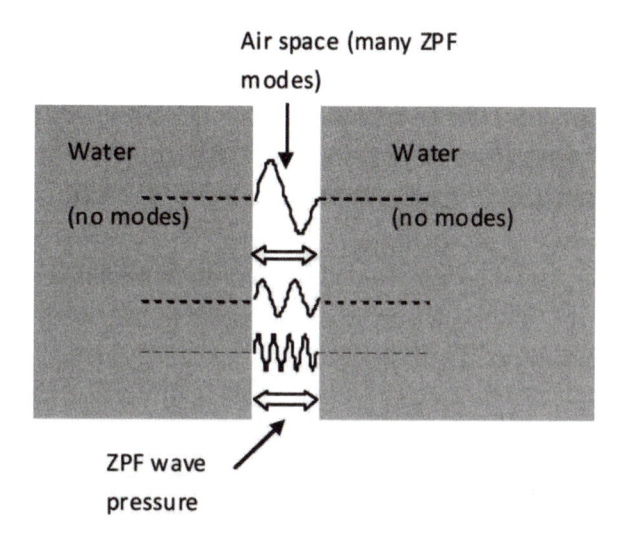

Trapped ZPF between layers of water allows anti-bubbles to exist.

The main difference between the ZPF and the white light of a hot furnace is that it includes a large component at zero frequency. This component is apparently made up of loops of magnetic and electric flux. The vacuum, if it could be seen, looks like the inside of a box of Cheerios with magnetic flux loops lying in every direction, unseen and undetectable. Each loop carries a quantum of magnetic flux. The flux loops can be as small as the Planck length, and are thus minuscule next to the size of the particles. Around each particle the flux loops form a sphere made of flat loops. The

Beyond Einstein: GEM Unification

electric field of the particle is mostly normal to the magnetic lines of force in the loops but because they are locally flat a small amount of the Poynting vector ExB force faces inward. Summed over the sphere the Poynting vectors add up to make gravity. Thus, the gravity is due to the electric charge of the particles interacting with the magnetic loops making up the ZPF. At larger scales the electric fields of the particles cancel each other, but the Poynting vectors add and so the mass of particles of canceling electric fields add to each others' Poynting flow, leading to gravity on large scales. Because of the nature of the identical magnetic flux loops filling space the field intensity of the ZPF is nearly uniform and thus spacetime is essentially flat.

Anti-bubbles in water, a thin layer of air separates two water surfaces. What force keeps the surfaces apart? The author has hypothesized that the ZPF is responsible.

But what is the nature of the particles? Einstein knew that no theory of the unification of gravity and EM would be complete without the theory of particles. Einstein, in his last published papers was able to show that in regions of high

Beyond Einstein: GEM Unification

spacetime curvature, particles would themselves follow geodesic trajectories assigned to masses. The particles are essentially the hidden dimension. If we imagine the original state of the cosmos as pure vacuum, with a ZPF field in place due to EM fields at the Planck scale, then the vacuum has energy but no particles. The only meaningful length scale is the Planck length. The Hilbert action principle over this four-dimensional emptiness gives us an Einsteinian vacuum with gravity waves but no matter. Without charged particles the character of the ZPF as EM fields is meaningless, since no charged particles exist to move under the influence of the EM fields. However, this empty vacuum is pristine and orderly. But now comes something to disrupt this order and lead to entropy.

We can understand this by looking into the Planck scale. We can perform a "Gedanken," or "thought" experiment, by placing a single atom of hydrogen, a proton and an electron, into a box and squeezing it. The Heisenberg uncertainty principle ensures that, as we squeeze the atom, it is like compressing a gas, the electron and proton become hotter as they are compressed into less space. Thus, the electron and proton will begin to rise in energy, and the atom will ionize into a plasma. Finally the electron and proton will gain in mass as they are squeezed further and their motion becomes relativistic. Ultimately, the box will gain so much mass as it is squeezed to the size of Planck length it will become a black hole. However, it is so small that once it does become a black hole it will undergo Hawking evaporation into a shower of particles and antiparticles. The baryon number of the proton and the lepton number of the electron have been destroyed and the proton and electron pair is now indistinguishable from a vacuum. This means at the Planck scale the lepton and baryon numbers become canceling quantum numbers. We can therefore, create a model where the Kaluza-Klein fifth dimension, when it appears and deploys, generates the baryon and lepton quantum numbers.

A hidden fifth dimension appears, starting in size at the Planck scale and growing to monstrous size in terms of the Planck scale until it stops. It captures some of the electric and magnetic flux from the Planck scale forming a spinning charge. The hidden dimension viewed in the Planck scale is not hidden at all. It is enormous, yet unlike the other limitless dimensions of height, breadth, and width the new fifth dimension is limited in size from our perspective. Viewed from the Planck scale the fifth dimension inflates or deploys to enormous size then stops. This size is the mesoscale size of 43 times smaller than an electron. So from our perspective it is too small to detect directly. When the fifth dimension deploys it creates an EM field separate from the gravity fields. Suddenly electric and magnetic fields can exist at a scale outside the Planck scale, because some of the EM fields that existed at the Planck length are captured and bound to the fifth dimension when it deploys. When the fifth dimension of Kaluza and Klein appears, so does light and so does hydrogen. Edward P. Lee of Lawrence Berkeley Laboratory first suggested this correlation to the author.

To understand the universe to useful approximation, we must understand the atoms as well as the void. Here we make a crucial approximation: we imagine that the universe began as pure hydrogen, a plasma of electron and protons with no neutrons. This is only an approximation. It is estimated that 1 in 6 baryons in the early universe was a neutron, this is required to give the observed distribution of light elements in intergalactic space. However, a pure hydrogen universe is a good start. This *Two Particle Paradigm* was also used by Einstein for the same reasons: it separates the two long-range forces of nature EM and

Beyond Einstein: GEM Unification

gravity, from the short range forces, weak and strong. This approximation creates a cosmic problem that can be solved.

The size of the hidden dimension is measured by the classical radius of a charged particle with both the electron charge and a mass that is the geometric mean of the rest masses of the proton and electron. Its size depends only on the charge and masses of the electron and proton. It is a new thing, yet born out of the vacuum. The quantum vacuum in turn, has the Planck length, which depends only on the vacuum quantities of Planck's constant, the speed of light, and Newton's Gravitation Constant.

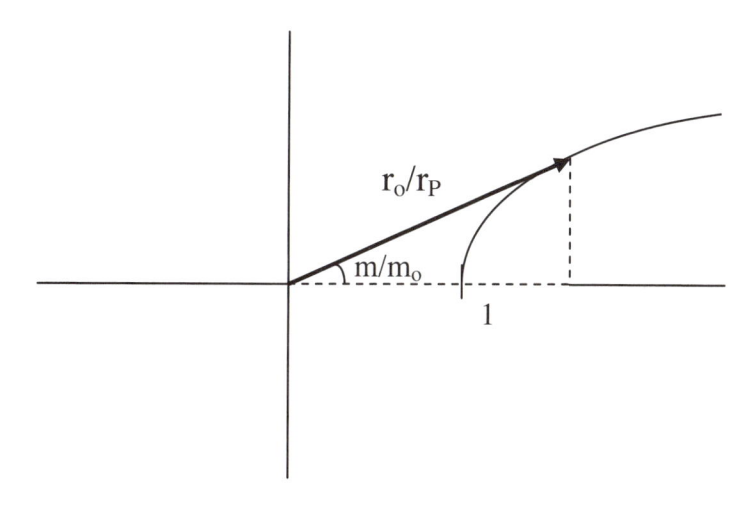

A method of visualizing the exponential dependence of the size of the hidden dimension on the ratio of the proton mass to a union particle mass. As the mass ratio grows, so does the hidden dimension grow from the Planck length. Therefore, the electron-proton pair and EM and gravity appear as separate forces at the same time.

Beyond Einstein: GEM Unification

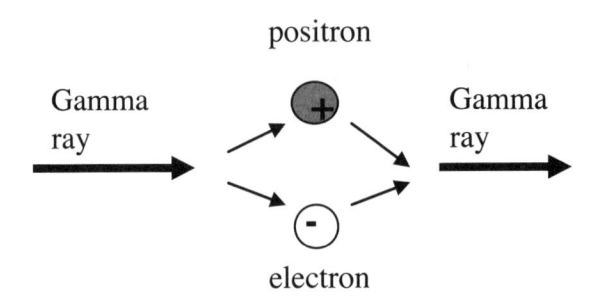

```
                    positron
     Gamma            (+)          Gamma
     ray            ↗    ↘         ray
     ───────→                      ───────→
                   ↘    ↗
                    (-)
                  electron
```

Ordinary virtual pair creation in the vacuum where the electron-anti-electron (positron) pair appears from a gamma ray in the ZPF then recombines and annihilates to form a new gamma ray.

What also appears out of the vacuum are particles that do not annihilate each other. The fifth dimension does double duty; it allows EM fields to separate from gravity but it also allows two types of particles to occur. It generates electrons and protons. At the Planck scale particles and antiparticles of identical mass and opposite charge constantly appear, then annihilate each other to form a vacuum again. A four-dimensional vacuum with a ZPF is just that despite its population of virtual particles. However, when the fifth dimension appears it allows two different types of particles to occur that cannot annihilate each other despite being of opposite charge. One is light and one is heavy. Their charges cancel but their substance cannot.

The electron and proton particles can both be considered fundamental because they do not decay. However, the proton is made of three quarks and is thus more complex than the electron. Basically, it can be seen that the triune character of the proton and unitary character of the electron reflects the

Beyond Einstein: GEM Unification

fundamental structural asymmetry of space and time. If we concentrate on the electron we can understand that we can model it not as a point, but as a spherical shell of charge held together by gravity. This occurs because for a point charge to have finite mass the electric field around it must saturate, that is, reach a maximum and then no longer increase. This occurs because in the GEM theory electrostatic potentials near a particle can act like gravity potentials and satisfy the same equation as gravity. This solution for the potential is found using the same equation as the Schwartzchild solution for the black hole but has a positive sign instead of a negative sign. Instead of a black hole, the solution forms a white hole. In short, as the shell shrinks, the electric field saturates at the charged shell's surface, but gravity does not.

The quantum mechanical vacuum must also enter here, as we shrink the charged-shell electron. The effect of quantum mechanics means that the charge of the electron we see at normal scales is only a diminished value that reflects the partial neutralization due to the quantum vacuum. At the deep subatomic scale the "bare" charge emerges, and this charge appears to be approximately 12 times the charge seen at normal scales. What basically happens, in the GEM theory, is that near the Planck scale, the electron charge becomes a unit of action, and reverts to Planck's unit of action: Planck's constant. So the electron actually gets harder to compress as one approaches the Planck scale due to this failure of quantum charge neutralization. However, even this increase of charge effect saturates, but gravity continues to increase. Finally a balance point occurs, and with a bare charge it produces an equilibrium radius of the Planck length. If we add spin to the electron, it will be too stable to collapse into a black hole, so it must spin to exist. Therefore, though it describes a model that can exist only at the abstract Planck scale, the model of an electron, and by extension the quarks, as particles held together by gravity and stabilized by spin is not unphysical. However, at the Planck scale the physical quantity we associate with sub-atomic particles, the electron charge, disappears and is replaced by Planck's constant. This is called the 'Planck charge.' Thus, at the Planck scale everything is very simple, with only a few physical constants: G, the speed of light and Planck's constant. The quantities we see at our scale, the masses of the proton and electron, and their charges, are not seen.

Therefore, Einstein's concept that electrons, and by extension, the quarks, are held together by gravity because of the saturation of electric fields near the Planck scale was basically correct. The electric field becomes saturated because near the particles the electrostatic potential begins to behave like a metric and follows the same equation, only instead of a black hole metric, it is a white hole metric. The saturation of the electric field can be considered as a change of dimensionality near the Planck length from three dimensions to one, with electric and magnetic flux forming bundles or strings. In a string universe, the electric force between two charges is constant, just like the saturated electric field at the surface of our model electron.

Beyond Einstein: GEM Unification

Electric repulsion

Gravity attraction

Particle shell of space charge

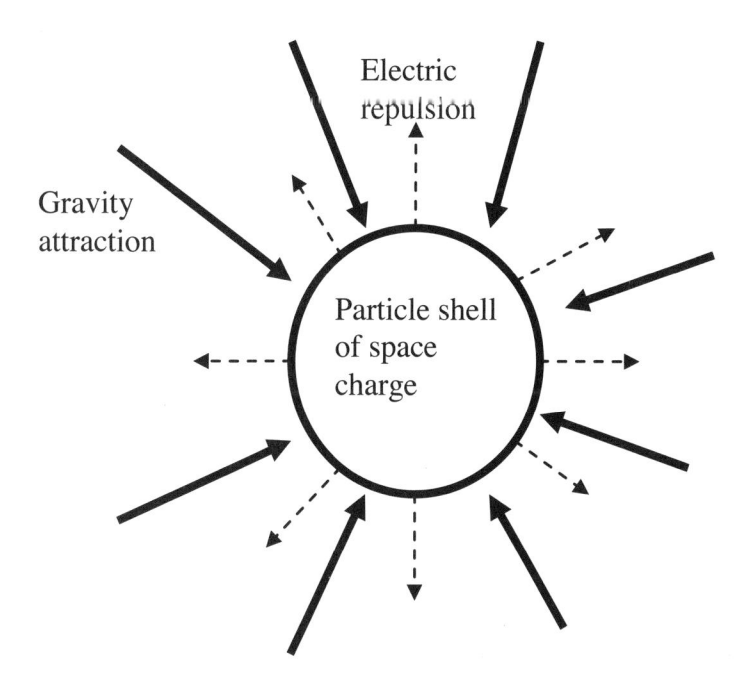

A conceptual model of an electron held together by gravity.

Beyond Einstein: GEM Unification

Field Strength vs Radius

- Gravity field
- saturation
- Electric field

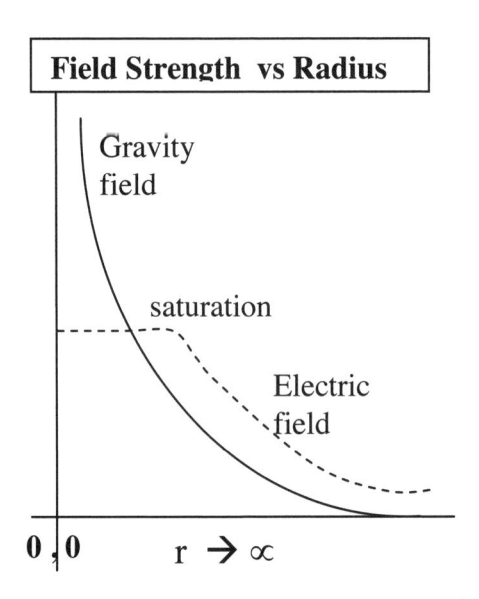

0,0   r → ∞

The strengths of electric and gravity fields near the surface of an electron or other charged particle.

This string potential, or "white hole" metric model of electric and magnetic fields near the Planck length shows that the fabric of space-time at this level can be modeled as a fibrous network of bundles of magnetic and electric flux. The universe at this level looks like a foam of electric and magnetic fields. This is similar to concepts in "loop gravity" championed by Lee Smolin of the Perimeter Institute (PI).

Beyond Einstein: GEM Unification

Lee Smolin.

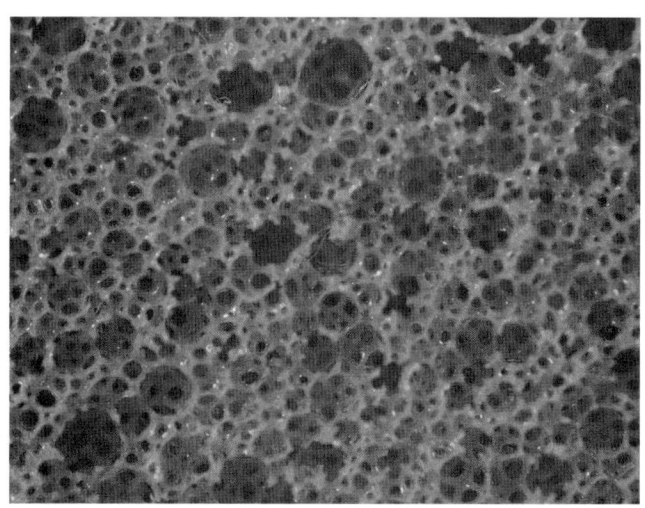

Plastic foam magnified, representative of the EM structure of spacetime near the Planck length in the GEM theory.

Based on this model, the movement of particles through spacetime, observed near the Planck scale, no longer looks

225

Beyond Einstein: GEM Unification

like our familiar motion of a ball bearing or ping-pong ball. It consists of a diffusion of the wave function of the particle down multiple pathways at once. Instead of one ping-pong ball we would see a cloud of ping-pong balls bouncing down multiple channels at once. The aggregate motion of the particles would be that part of the wave-function finding the shortest path through spacetime subject to the forces it feels. The particle basically sees itself alone in a basically flat universe with certain small curvatures of spacetime, where all forces look like a curved space. Put differently, every particle, depending on its charge-to-mass ratio, sees a different universe. The collaboration of large numbers of particles, a cloud of hydrogen for example, determines what we experience as spacetime.

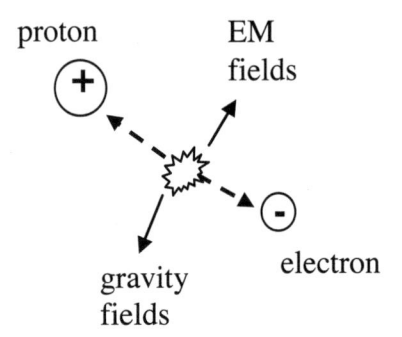

Gravity and EM fields separate and become distinct long-range fields as a proton and electron pair separate from the vacuum and become separate particles, both events triggered by the birth of the fifth dimension.

Therefore, the birth of the particles results in a heavy particle and light particle of opposite charge. These are the electron and the proton that make up hydrogen. They are the simplest members of the lepton and the baryon families,

Beyond Einstein: GEM Unification

lepton meaning light, and baryon meaning heavy. They cannot recombine to annihilate back into the vacuum. This is the preference for matter over antimatter and the non-conservation of baryon and lepton numbers required by Sakharov for the universe to be created. The hidden dimension is limited in extent in human terms, but it does have an up or down, a positive or negative. The positive and negative are measured by charge. The "up-ness" is the mass being large and the "down-ness" is the mass shrinking to that of an electron.

Therefore, the universe began with a four-dimensional vacuum with gravity as the only physical force. Then a fifth dimension appeared. With it appeared light, separate EM fields, and protons and electrons, making hydrogen. So the universe started in a vacuum with gravity waves and ended with two particles, electrons and protons, and, acting between them, two force fields, gravity and EM. From the Planck scale, with only G, the speed of light and Planck's constant, appears a new scale, the mesoscale, with new constants, the electron charge, the proton and electron masses. The number σ, 42.8503... is embodied in this new scale, by which we exist. This number is intertwined, as it turns out, with the number α or 1/137 so familiar in quantum mechanics, which is the ratio of electric charge to the Planck quantum of action: h. So sigma and alpha are related; one determines the other. This transcendent truth underlies reality as we know it. The equation that relates sigma and alpha is called a transcendental equation, so once again the language of mathematics points us to poetic truth.

From the Planck length the length of the hidden fifth dimension grows and then stops, forming the basis of the subatomic scale. The exact processes by which these things occur are details that escape us for now. The idea of a fifth dimension that suddenly appears, deploys, and stops at a

Beyond Einstein: GEM Unification

given size is an idea that makes sense in everyday experience. A mushroom sprouts, a bird's egg hatches, a flower blooms. They grow to a certain size, , then they stop. That is enough for us right now. Heisenberg and Bohr were right in the sense that much of the subatomic scale is a mystery that is far removed from our everyday concepts, but Einstein was also right in that much of the everyday helps us understand the abstract, for they are all of the same cosmic fabric.

Thus, the GEM theory at its most basic formulation is the following: Gravity fields are due to the microscopic action of crossed electric and magnetic fields making up spacetime's fabric. The universe came to be with the birth of the fifth dimension, that being added to four dimensional spacetime allowed EM fields to break free of the Planck scale and become distinct fields separate from gravity, and simultaneously allowed the vacuum to split into two particles of equal charge but different masses. Thus, we began with a vacuum but end up with a cosmos full of hydrogen with two long-range forces, EM and gravity, to rule its dynamics.

Much complexity is covered up in this summary. The particles occur because four spheres of collapsing or expanding waves suddenly get short circuited by the fifth dimension with the result that time independent spheres appear that are either space-like with three subdimensions and the fifth dimension taking the place of time, or time-like particles with the fifth dimension taking the place of radius. Both are spheres and isotropic in space, but one has three subdimensions and is heavy and the other is unitary, like time itself, and is lighter.

The proton and electron can be viewed as compact versions of the two parts of four-dimensional spacetime, expressed in charge. The proton is a compact image of a sphere expressed in charge, with three inseparable but distinct subdimensions. The proton is triune, consisting of three internal quarks, yet indivisible. The electron is, however, a compact image of time expressed as a spherical single point charge. It is unitary and one dimensional.

Looked at another way the proton can be seen as a hot particle, with three sub-charges thrashing around like three molecules in a tight box. Because it consists of three internal particles, it has entropy. If the proton had two internal particles they could orbit each other smoothly; three particles are disorderly and asymmetrical. The proton is the source particle for gravity fields of the universe. Gravity can be considered to be closely related to hot EM fields full of modes and a spectrum of frequencies. EM fields on the other hand are mostly cold, having a narrow range of frequencies and being driven by the dynamics of the electron.

The proton can be modeled as a ball of hot light, like a furnace, whereas the electron can be modeled as a simple sphere carrying a static electric charge on its outside.

Therefore, we can see that the universe came to be because the vacuum was too ordered and constrained. This apparently displeased God. The fifth dimension seems to have been born so the cosmos could increase entropy. The cosmos needed to evolve.

How do we know this theory is a good approximation to reality? First, we can test the model of gravity as an ExB or Poynting drift using a computer simulation. When this is done we see that the electron and proton do a strange dance as they accelerate. The cycloidal motion they exhibit becomes smaller as the magnetic field increases, but the acceleration stays the same as long as the electric field increases proportionally.

Beyond Einstein: GEM Unification

The most important test of the GEM theory is whether it predicts the value of the Newton gravitation constant relative to the masses of the proton and electron and the charge of the electron. This can be seen easily by a model that correlates the ratio of proton to electron mass with the size of the hidden dimension relative to the Planck length. In the Kaluza-Klein theory the square root of the masses of the particles is important as opposed to the masses. If we use a simple model where the square root of the ratio of the mass of the proton to the electron is proportional to the logarithm of the size of the hidden dimension to the Planck length, we find this model is very successful. The masses of the proton and the electron separate from a common mass and become different as the hidden dimension deploys. This means that the cosmos began with a particle and antiparticle pair of opposing charge which then separated smoothly to become the electron and proton. This model gives us the signature equation for the GEM theory:

Logarithm of the fifth dimension size measured in Planck lengths = the square root of the proton-electron mass ratio

The equation says that the quantity 42.8503 ... , which we will call sigma, σ, is the most important number in the simple universe we have created. As in the whimsical sci-fi novel, *Hitchhikers Guide to the Galaxy*, it is the answer to the *great question of how gravity and EM forces and protons and electrons are unified*. Alternately, it is the number that underlies the creation of the universe consisting of hydrogen and having two long-range forces, EM and gravity. We can invert this equation (see chapter notes) and obtain an expression for G that is the Newton gravitation constant

Beyond Einstein: GEM Unification

The Newton gravitation constant expression is accurate to one part in one thousand. No other derived expression for the Newton gravitation constant exists in terms of other physical constants that even approaches the accuracy of this expression. It was first derived by the author and published in this form in 1988, it has now been improved.

The expression for the all important gravity constant G tells us the mechanics of the gravity force. As was hypothesized by Andrei Sakharov, it is a force mediated by EM photons, like EM forces themselves. The quantum mechanical probability alpha, α, gives the probability that an electron charge will emit and absorb a photon. However, for the gravity process the probability α is reduced by an enormous factor having to do with the ratio of the size of the cross section of the particles of the hidden dimension size to the Planck length cross section. The particle cross section for the absorption or emission of GEM photons is basically the Planck length squared, whereas, for EM process photons it is the Thompson cross section of the EM classical radius squared. That is, all the particles look the size of Planck length squared to GEM photons, but to EM photons they look much larger. M.J. Clark has proposed similar models.

Therefore, Einstein was right in his basic hypothesis that EM and gravity were fundamentally related and could be unified with a useful outcome. He was also right that this unification had to involve the unification of the electron and proton. He was also right in his conjecture that models of the electrons and quarks could be found where the charged particles were extended objects held together by gravity, though the scale of this effect is at the Planck scale and thus difficult to test. As is pointed out by Brian Greene in his excellent book on string theory, *The Elegant Universe*, the idea that particles are not points but extended objects, is the

Beyond Einstein: GEM Unification

heart of string theory, which is presently the main stream of theoretical thought in physics.

The GEM theory uses the very tools Einstein created and endorsed for physics: the theory and formalisms of special and general relativity, the ZPF, the photoelectric effect, and the theory of the hidden dimension of Kaluza and Klein. Unfortunately, he rejected all the quantum mechanical concepts that he himself had helped to develop in his own unification quest. However, as was explored in the movie *Insignificance*, he may have come to regard the end result of his quest with the same ambivalence that afflicted Heisenberg in his quest to deliver the atomic bomb to Hitler. To unify the fields would not be without consequence.

Einstein's first great theory, special relativity, led to the destruction of two cities in Japan, a nation he loved. This saddened him greatly. Perhaps his concern that a unified field theory would lead to even greater destruction robbed him, as it did Heisenberg, of the intellectual drive he needed to assemble the pieces of the puzzle that he had in front of him. As it is, Einstein formed the basic hypothesis of EM-gravity unification for the physics community, and provided a kit of components and tools that would lead to GEM unification.

Therefore, the GEM theory gives us a physical model of gravity fields as EM phenomena and also gives us an expression for the Newton gravitation constant in terms of EM and quantum mechanical quantities. That is, it gives the reason gravity forces are so weak between subatomic particles compared to EM forces. Basically, the EM photons that mediate gravity are much less probable to be emitted or absorbed than regular EM photons, with the result that the force between individual particles is so much weaker.

Beyond Einstein: GEM Unification

No theory is any good unless it can be tested. How do we test the GEM theory? Initial tests are easy.

First, we note that the hidden dimension size, in a quantum mechanical vacuum, will attract a quantum to occupy it. The mass of such a particle will be 3000.6 MeV. This particle has been detected and is called the Sigma (3000) particle. It decays into a proton and an electron plus some mass-less particles. The particle lies between two mesons in mass, the charmed eta meson at 2980 MeV and the J/ψ at 3095 MeV, both of which figure large in tests of matter and antimatter preference in the universe. Both these particles decay into either proton-antiproton pairs or electron-positron pairs. Much activity occurs in this energy range that corresponds to the hidden dimension size. The author is grateful to Eric Davis of the Austin Institute for Advanced Study for suggesting this inquiry. The hidden dimension size seems to function as a generic length where quantum fields can find resonance, as evidenced by further clusters of particles at higher energy harmonics: B mesons at approximately 6000MeV and an Upsilon meson at approximately 9000MeV along with baryons at similar energies. So there is evidence of generic-geometric hidden dimension behavior in particle mass-energies.

Second, we can invert the model of the proton and electron mass ratio being related to the size of the hidden dimension as it deploys, giving us a formula for the Newton gravitation constant. We find that the GEM theory predicts the Newton gravitation constant to high accuracy, within a part per thousand (see chapter notes) .

Finally, there is the fabled number 42.8503..., which is approximately 43. This number pops up in the periodic table as the atomic number of the element Technetium, which, though surrounded by stable elements, has no stable isotopes. This is counterintuitive, for one would expect a

Beyond Einstein: GEM Unification

nucleus with 43 charges, in the GEM theory, to be exceptionally stable, but instead, the opposite is true.

We are now ready to discuss the main technological consequences of this theory: gravity modification and antigravity.

vanadium 23 **V** 50.942	chromium 24 **Cr** 51.996	manganese 25 **Mn** 54.938	iron 26 **Fe** 55.845	cobalt 27 **Co** 58.933
niobium 41 **Nb** 92.906	molybdenum 42 **Mo** 95.94	technetium 43 **Tc** [98]	ruthenium 44 **Ru** 101.07	rhodium 45 **Rh** 102.91
tantalum 73 **Ta** 180.95	tungsten 74 **W** 183.84	rhenium 75 **Re** 186.21	osmium 76 **Os** 190.23	iridium 77 **Ir** 192.22

Technetium: of atomic number 43, sits in the middle of stable elements, yet has no stable isotopes.

Chapter 11: Antigravity and Human Flight

So, if gravity and EM fields are unified, and are thus manifestations of the same underlying phenomenon, the quantum Poynting force, then the question becomes: can this be used for practical purposes? The answer, apparently, is yes. It is as if the cosmos was made to be user friendly.

Gravity is the result of actions of microscopic EM fields and waves on the charged particles that make up all matter, the electrons and quarks. EM fields created by technology have much larger size. How is a phenomenon that relies on the microscopic scale going to be affected by large scale phenomena? The answer is seen by analogy with another technology in everyday use: the airplane.

The airplane flies because of the Bernoulli effect, which says that static pressure, the pressure that fills a balloon, and dynamic pressure, the force per unit area of wind, are related so that their sum is always constant. The airplane has wings that curve on the top and are flat on the bottom. This means air moving over the wings goes faster over the top than the bottom like water in a brook moves faster over a stone. Because the air moves faster, its dynamic pressure is higher; hence, the static pressure is lower.

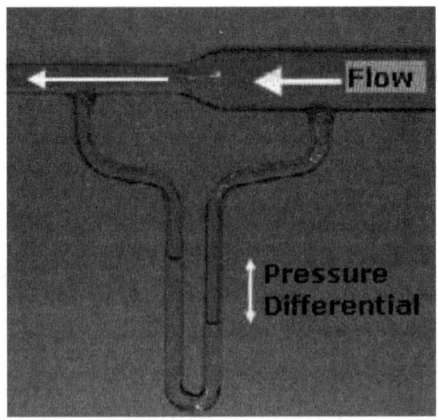

Bernoulli effect causing lower pressure with more rapid flow in a smaller pipe.

The static pressure of air is due to innumerable molecules and atoms moving and bouncing off surfaces. This bouncing adds up to a force per unit area, the sum of microscopic effects. Dynamic pressure, however, is due to the mass movement of the same molecules in large-scale currents. Yet nature treats these two effects, a microscopic effect and a large scale effect, the same.

No more magnificent example of the Bernoulli effect can be found than in the tornado, where the vortex motion of the wind at high speed is compensated for by low thermodynamic pressure. The spinning winds at high speed would fly apart because of centrifugal force; however, the low pressure of the air at the center of the vortex presses the winds together. Therefore, it can form a balanced and stable entity despite tearing apart everything it encounters. Thus, the tornado exists, and can exert tremendous force, because the vortex can exist as a balanced entity within itself. Most people do not realize that the same vortex principle underlies the action of an airplane wing.

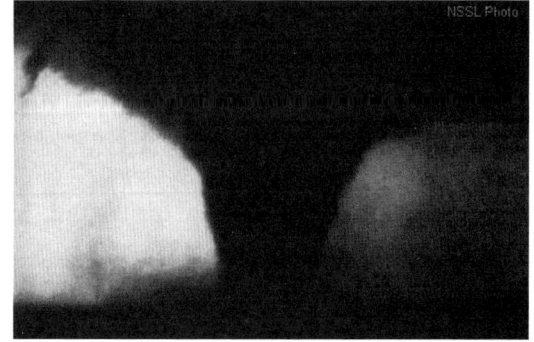

Tornado funnel cloud.

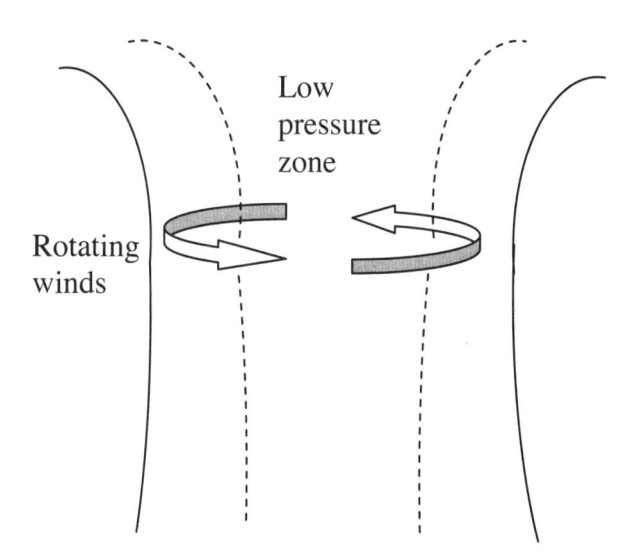

A basic diagram of a tornado or dust devil showing a low pressure core.

An airplane flies because the wings hold it up in the air. The wings in turn are essentially two plates that feel a net force due to the higher pressure on their bottom surfaces.

Beyond Einstein: GEM Unification

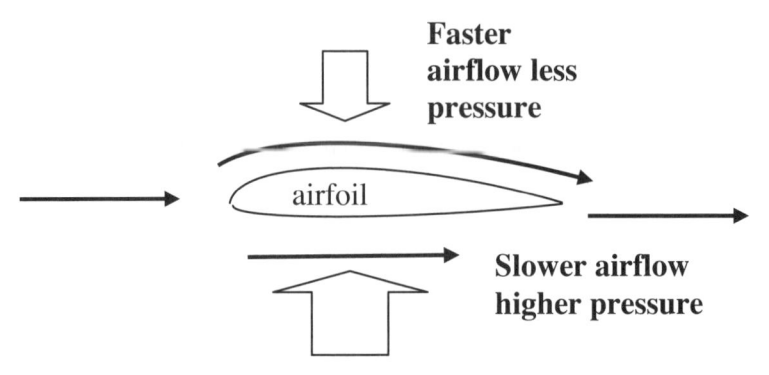

An airfoil designed to use the Bernoulli effect to create lift.

The air flows more rapidly over the top surface of the wings; according to the Bernoulli equation the pressure there is lower. The sail on a sail boat works the same way when it tacks upwind. However, there is another way to look at this problem that was discovered in Germany in World War I, the fact that the difference in flow above and below the wing surfaces creates a vortex flow pattern around the wing as it flies. In Germany it was discovered that a spinning cylinder entraining a vortex of wind around it by friction, functioned as a wing in an airflow. The rotating cylinder was not practical as an airplane wing, being too heavy and complicated. The spinning cylinder has been used as a sail on ships, where it functions the same way. However, the vortex principle of the wing is now the basis for wing design. The vortex pattern around the wing can be seen through the superposition of the flows, where vortex motion adds to the flow at the top of the wing, increasing the speed of the flow and subtracting on the bottom to lessen the speed of the flow. The net effect is magnified because the square of the flow speeds, not the speed itself, creates the pressure difference. The airplane thus flies because of the vortex that forms around the wings.

Beyond Einstein: GEM Unification

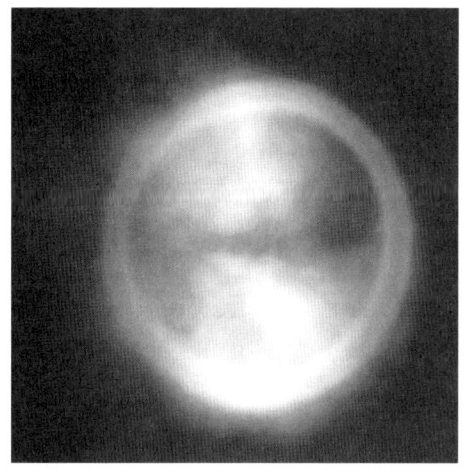

A ring vortex : a smoke ring.

A vortex produced by a aircraft wing, made visible by smoke.

In a hovercraft, the vortex principle is used indirectly to keep air pressure corralled in a cushion beneath the hovercraft. The vortex ring, commonly seen as a smoke ring, serves as a barrier to help keep high pressure air confined. In a smoke ring vortex, the vortex lines of force run a circle around the ring. One could say that the vortex lines form a

239

Beyond Einstein: GEM Unification

vortex of their own; however, the effect is merely to create a barrier confining the pressure at the center of the ring that will lift the hover craft. The region of high pressure air presses on the bottom of the hovercraft and floats it above the ground. However, if the hovercraft attempts to rise too far above the ground not even the ring vortex can keep the column of high pressure air contained and the hovercraft loses lifting force. This requirement to be near the ground is called "ground effect" and limits the hovercraft to fly close to the ground. The hovercraft needs the ground itself to help confine the pressure beneath it to keep it aloft. For this reason conventional hovercraft are restricted to flying close to the ground over relatively smooth surfaces. However, circular wing craft have now been successfully developed.

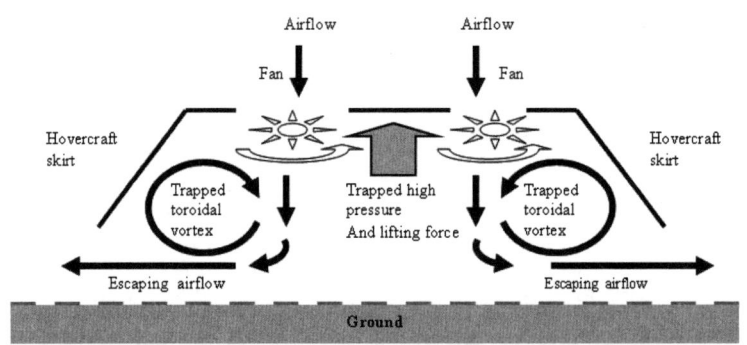

Hovercraft diagram, showing trapped toroidal vortex.

Circular wing craft were apparently first investigated by the Germans in World War II. Like many strange German ideas investigated, the circular wing was part of their desperate search for a "wonder weapon," that could win the war against the allies. This research accomplished little for the Germans during the war except to dissipate precious war resources, a good thing for the rest of the world. However, nowadays, given new light weight materials and better

Beyond Einstein: GEM Unification

understanding of airflows, the circular wing aircraft is now practical, at least as an unmanned craft. These vehicles consist of a fan or turbine pulling air in from the top of the craft and blowing it downward. The torque of the fan is compensated for by making it a counter rotating pair or having grooves cut into the upper surface to create a small counter vortex flow. The downdraft generated by the fan or turbine creates a rocket-like downward thrust; however, this is not where the majority of the thrust is generated to lift these craft. Most of the thrust for a successful circular wing craft comes from the Bernoulli effect on the top surface of the craft, which functions as a wing. The airflow in such a craft is pulled across the upper wing surface to create the asymmetry of flow seen on an ordinary airplane wing. The top surface of the circular wing craft is thus the "airfoil" that generates most of the lift.

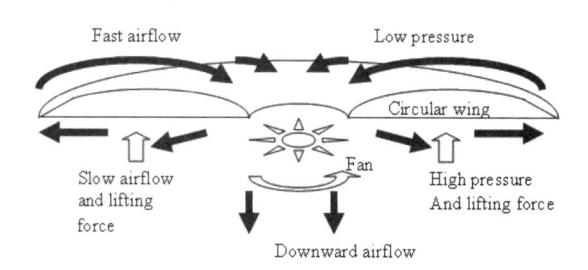

A diagram of a circular wing aircraft.

Such craft were first built in the 1950s in the U.S. Army's "Pawnee" program. This program was ultimately successful in creating a flying circular wing craft that could rise out of its own "ground effect." The resulting craft was not successful in replacing the helicopter, where a small lightweight wing is simply rotated to generate lift while the craft is stationary. The circular wing craft is now usually employed as a UAV (Unmanned Air Vehicle) because it requires tighter engineering margins than a helicopter or conventional airplane. This illustrates that the same physical effect, the Bernoulli effect, can be employed many ways to create many types of aircraft, with many roles, but the engineering details of such employments varies greatly in difficulty. It will be the same with antigravity technology.

The Hiller VZ-1 Pawnee experimental circular wing craft .

Beyond Einstein: GEM Unification

Micro-Air Vehicle using circular wing principle.

Avro Canada VZ-9 Avrocar circular wing craft.

Antigravity technology relies on the "Vacuum Bernoulli Effect" that emerges from the GEM unification theory. In the GEM theory we have a "Vacuum Bernoulli Equation"

Beyond Einstein: GEM Unification

that is the same as the Hilbert action that is minimized by the combination of Einstein and Maxwell's equations. The Vacuum Bernoulli Effect takes its name from its close analogy with the aerodynamic Bernoulli effect, which gives us human flight. The Vacuum Bernoulli effect will give us, once it is translated into engineering, an even greater variety of flight vehicles using antigravity. To understand this effect, one must understand how gravity creates a pressure similar to air pressure.

Gravity pressure is an essential concept in understanding antigravity. Gravity is normally thought of as a force between objects like stars or planets that occurs between their centers of mass. For a star or planet, being spheres, this is a simple way to think of the gravity force between them, because, as Newton proved, the force can be calculated as if the masses were concentrated in their centers. What of the forces within stars or planets themselves? To understand the internal forces in a star or planet, particularly the forces that make them a spherical shape, one must understand the concept of gravity field pressure.

Gravity fields as pressure in space is described mathematically as the gravity field intensity squared, divided by the all important Newton gravity constant. The idea that this expression for gravity pressure uses units of thermodynamic pressure was used by Sakharov in his breakthrough analysis equating gravity to a radiation pressure from the ZPF. Here on Earth, the gravity pressure merely works tirelessly to contain the pressure of ordinary substances such as air, water, and rock.

Earth is composed of rock, and at high temperatures and pressures the rock melts to form a liquid magma, often full of dissolved gas. The solid cool crust of Earth then floats on this sea of liquid magma that makes up Earth's bulk. When a volcano erupts, we can see that it is venting high pressure

Beyond Einstein: GEM Unification

from inside the earth. The earth is then full of high pressure molten rock. What keeps it from exploding completely, instead of having an occasional local volcanic eruption? The answer is that gravity field pressure confines the pressure of the molten rock in the earth.

Force is created by pressure imbalance, not pressure itself. The total pressure at any point in space, when calculated mathematically, is due to the thermodynamic pressure of molten rock, air, and water, but must also contain the gravity pressure. This mathematically complete pressure moves things when imbalances occur and is constant in space when things are in equilibrium. To illustrate this one can hold a pressure gauge in one hand and a gravity gage in another at Earth's surface and note that, unless one is sitting over an erupting volcano, the pressure of the rock making up the earth is zero. The gravity of the Earth is high when calculated as a thermodynamic pressure. The gravity field pressure at Earth's surface is millions of atmospheres of pressure. We do not feel this pressure directly, only its differences, as our weight. Now, let us take an imaginary journey, like ghosts, through the rock of Earth itself. Since this is an imaginary journey we can bring along our imaginary gravity gauge and our imaginary pressure gauge. We discover a marvelous thing when we journey ghost-like through the rock: The pressure of the rock around us gets larger and larger as we go deeper, but gravity field pressure gets weaker. Finally, we end our journey in the fiery heart of Earth itself. Here, the pressure of the rock around us, supporting the weight of the whole Earth, is millions of atmospheres, but gravity is zero, as is its field pressure. Gravity is zero in the center of Earth because Earth is a sphere and in its exact center the mass of gravity pulls everywhere equally so the net gravity force is zero. Thus we journeyed from the place on Earth's surface, where gravity

Beyond Einstein: GEM Unification

pressure was millions of tons per square yard and rock pressure was zero, to the center of Earth where the gravity pressure is zero but the rock pressure is millions of tons per square yard. The two pressures changed places when we moved from the center to the surface and vice versa. Also, if we check our records of gravity and pressure, we find the total mathematical sum of the pressures was constant as we journeyed inward. So the total pressure, correctly calculated, includes the gravity pressure. So gravity is a pressure, like thermodynamic pressure. Now, how can we use gravity pressure to fly?

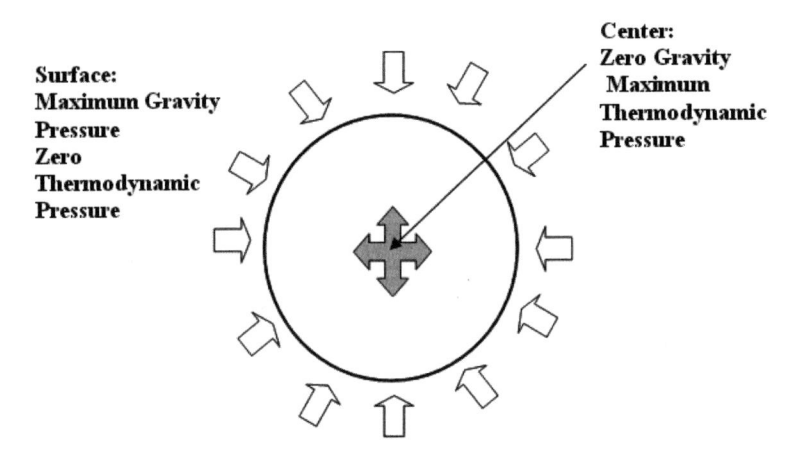

Gravity pressure creating a balancing thermodynamic pressure in a planet or star.

To harness gravity pressure to fly, we use the vacuum Bernoulli effect to create an imbalance of gravity pressure that will create a lifting force. The vacuum Bernoulli equation is different from the aerodynamic Bernoulli equation in that it includes a negative sign. The ordinary Bernoulli effect says that the sum of dynamic pressure caused by air flow and the thermodynamic or "static"

Beyond Einstein: GEM Unification

pressure must remain constant; that is, faster air equals lower air pressure. In the vacuum Bernoulli equation it is the difference, not the sum, of the gravity pressure, the "GEM static pressure" and the Poynting flow pressure, or radiation pressure, that is preserved as a constant. That is, more Poynting flow, more gravity pressure. It is as if the fluid aether has negative mass density so that its dynamic pressure when it moves is negative.

Stepping back into more conventional thought for a moment, one would ask, if we existed in a sea of negative mass aether, rather than air, and this sea of negative mass aether nonetheless followed the Bernoulli equation, what would be the most straightforward flying machine we could build?

Consider, first, that since the aether has negative mass density, so that a negative sign is introduced to the Bernoulli effect equation. This means that in an aether tornado, the pressure is higher at the center rather than lower. This means that houses implode rather than explode when the aether tornado sweeps over them. Perhaps then, to create high pressure underneath our craft we must create a confined vortex. The most straightforward, most simple, aircraft we can make is a hovercraft-like airplane.

We would want to make a disc-shaped hovercraft, and confine a vortex underneath it with a skirt and so create a region of high pressure. Instead of a set of vortex lines running in a circle around the bottom of the craft, like a conventional hovercraft, we will have a set of Poynting flow lines in a vortex flow pattern around the bottom. We will have a Tesla vortex.

EM foil

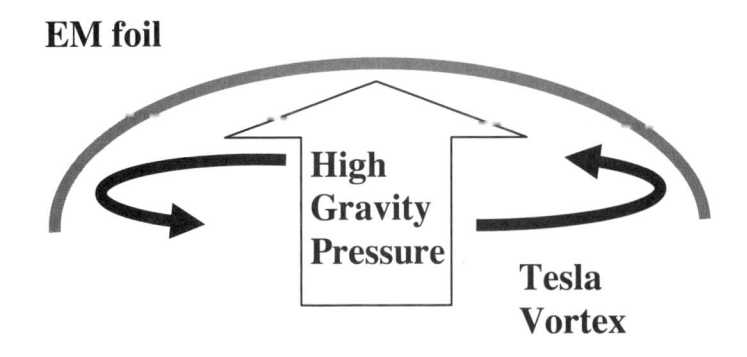

A Tesla vortex of Poynting vectors confined under a metal EM foil, by the vacuum Bernoulli equation, creates gravity pressure to cause a lifting force on the EM foil.

This Tesla vortex craft does not need to stay near the ground like aerodynamic hover craft, to keep the high pressure zone trapped beneath it. It needs no ground surface to confine it as in an aircushion hovercraft; it cares nothing for the ground. This craft maintains the high pressure zone underneath it, pressing on the bottom of the craft to create a lifting force because the high pressure is maintained by the Tesla vortex of Poynting flow. The gravity pressure underneath the craft is bound there immovably. This craft doesn't just hover close to the ground, it flies. It is a flying saucer.

To create the Poynting vortex underneath the GEM antigravity flying saucer, the three-phase power of Tesla is most practical. Three coils supplied by three-phase power is the most economical way to create a Poynting vortex. The three-phase rotating power activates three coils housed in blisters on the bottom of the dish. The dish should be called an "emfoil" because it shapes EM flow to create lift as an airfoil shapes air flow or a hydrofoil shapes water flow to create lift. The blisters covering the inductors for the three-phase Tesla vortex are necessary in air because the high fields required can ignite plasma in the air and this may cause the Tesla vortex to weaken or collapse, causing the craft to crash. For this same reason, flying through intense thunderstorms may not be advisable. Flying a GEM disk will probably require a skill set similar to that required by an airplane or helicopter pilot, and the same saying will apply: *"Any landing you can walk away from is a good landing."*

One can arrive at the same basic geometry for a GEM flying disk by designing a spacecraft that is propelled by pure electromagnetism and then trying to improve its performance. Such a craft can be built with only Maxwell's equations and uses the Poynting force. The craft will be a large parabolic dish with a light bulb or antenna as its focus. In space the craft can operate in two ways: It can reflect the light of the Sun as a solar sail, or it can make its own light. For the solar sail, the sunlight is absorbed and reradiated by the antenna light bulb and thus the light exerts both ingoing and outgoing pressure on the dish and propels the craft away from the Sun. Alternately, we can modify the dish geometry to simply make it a flat reflector. Far from the Sun, where the pressure of sunlight is no longer high enough to be useful, we can make the dish parabolic again and can turn on a nuclear power plant in a dome on the backside of the parabolic dish. The Poynting force caused by power flowing into the light bulb at the dish focus can propel the craft. Such an electromagnetically propelled craft can be built with present technology, but its thrust per unit of power expended is so tiny that only the solar sailing option is seriously considered. Only the unification of EM with gravity can improve the ship's performance, but the ship's basic form remains unchanged.

Beyond Einstein: GEM Unification

An electromagnetically propelled craft can be built using only Maxwell's equations and its shape is a satellite dish with perhaps a dome on top to house crew's quarters and a nuclear power plant. The reflector dish concentrates and shapes EM field for propulsion in this pure EM-propelled craft; only the details of the field inductor design under the dish are different from a GEM craft. A light bulb or antenna is used in the EM-only case and, minimally, three bumps in the case of the GEM craft.

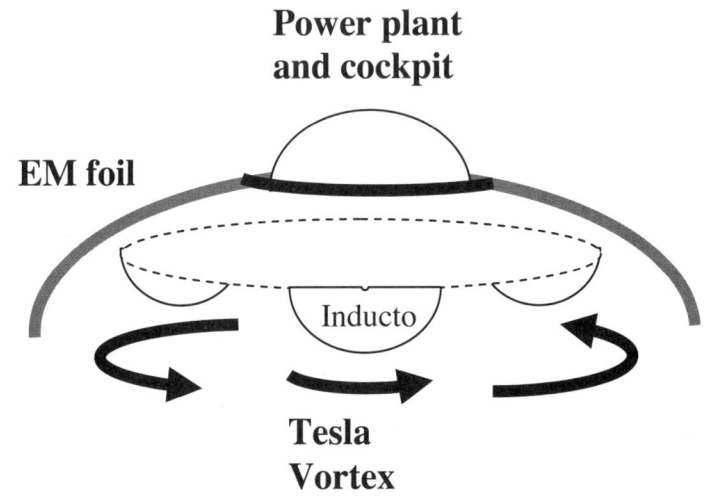

Power plant and cockpit

EM foil

Inducto

Tesla Vortex

A diagram showing: a cutaway side view of a GEM antigravity craft.

Beyond Einstein: GEM Unification

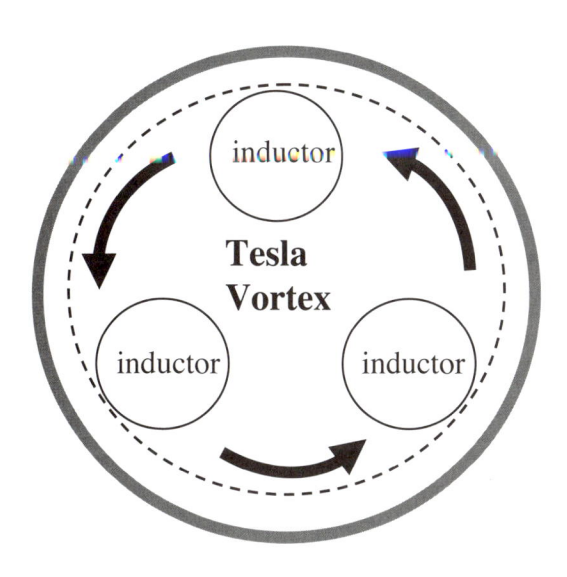

A diagram showing: a bottom view of a GEM antigravity craft with rotating Poytning vectors.

These discussions of the possibility of a GEM craft are supported by some recent experiments indicating modification of gravity by rotating EM fields. Japanese researchers Hasakya and Tacheuchi published research showing weight loss in metal disks spun by three-phase power in 1989. Their results echoed reports by a Russian physicist named Kozyrev. The author attempted to duplicate the experiments with as much accuracy as possible and saw similar effects using an EM-driven gyro without difficulty and published this result. A researcher ran a gyroscope with compressed air and saw nothing, reinforcing the opinion that this effect was electromagnetic in nature.

In the case of gravity modification by EM fields and its applications, it is possible that a government program on this subject exists that is deeply classified. This possibility has

Beyond Einstein: GEM Unification

been discussed in the excellent book *The Hunt for Zero Point* by Nick Cook. Such a program may have had its origins in German unified field research conducted in WWII , as has been proposed by Doctor Joe Farrell in his book *Secrets of the Unified Field: The Philadelphia Experiment, the Nazi Bell, and the Discarded Theory*.

If there were no ongoing UFO phenomena, featuring numerous reports of flying saucers and suggesting extraterrestrial visitation, the case could be made that the public had no "need to know" about such government programs. However, after reading the excellent books by Physicist Stanton Freidman such as *Flying Saucers and Science*, no intelligent person can rule out this possibility. Therefore, if the UFO activity represents extraterrestrial contact, then, since this is a democracy, the public must be informed of the basic facts of the situation if a proper government response is to be made. According to the U.S. Constitution, the people, through their elected representatives, run this country and determine its basic policies. They cannot do so if basic information about their situation in the cosmos is denied them. Extraterrestrial contact, if it has occurred, falls well within the realm of subjects that affect public policy, and the public should be given basic information. For instance, if we are maintaining a huge stockpile of nuclear warheads, despite the end of the of the Cold War, to deter an invasion from outer space, rather than some vague reasons having to do with nonproliferation, then this reason should be publically discussed. Unfortunately, some in the government may not see the situation in the same way, regarding an informed public as at best a nuisance, and at worst dangerous, to their programs. The scenario of a government cover-up of extraterrestrial contact has been discussed in many books,

Beyond Einstein: GEM Unification

and was explored by the author in a sci-fi novel *Morningstar Pass* under the pen name Victor Norgarde.

The GEM flying saucer is an obvious design for a craft operating near Earth in its intense gravity. Sightings of such craft are indications that someone may have harnessed GEM technology. Other craft of different shapes could probably be made using the Vacuum Bernoulli effect. The variety of craft that can be built will one day, it can be predicted, will exceed the variety of aircraft built today based on the aerodynamic Bernoulli effect. Think of it: the helicopter, the hovercraft, the Wright Brothers Flyer, the modern jet fighter, and the Space Shuttle, all rely on aerodynamic forces to fly. Like these various craft, the ultimate variety of GEM craft will be limitless. Also, the GEM craft does not only bend airflows to achieve lift and guidance forces, it bends spacetime itself. This means, according to the principle of equivalence, that gravitational mass and inertial mass must be the same, that reducing the gravity mass of a GEM craft so it floats must also reduce its inertia. Not only will a GEM craft float like a feather, it will have the same mass as a feather. This means that such a craft and its occupants can turn tight corners at amazing speed and feel no inertial effects. Ultimately, in deep space, it can be envisioned that GEM craft will not just bend space but twist it into a pretzel, or perhaps a wormhole.

To achieve "warp drive," that is, to travel faster than light, is the ultimate goal of any GEM program for human flight. Such a form of FTL (Faster Than Light) drive is necessary if the human race is to establish contact, trade, and cultural exchange with other intelligent species in other star systems. So it is necessary if the human race is to settle other living worlds that must lie also among the stars. Since the GEM theory allows control of spacetime geometry we can speculate that it will allow FTL. FTL phenomenon is probably not only possible, but it is probably exploited by the universe constantly. The most obvious way to accomplish FTL for human spaceflight is to modify spacetime geometry around a ship in space so that the rest of the cosmos thinks the ship is a tachyon.

A tachyon is a hypothetical particle that always travels faster than light. It has imaginary mass, making its direct interaction with particles in the cosmos impossible; however, such particles may make their presence known indirectly and subtly. The spacetime curvature an ordinary particle creates around itself in the void determines its mass as far as the rest of the universe is concerned. This suggests that FTL travel might be not only possible, but much easier that presently envisioned.

Present concepts for FTL travel involve collapsing a spinning mass the size of Jupiter and opening a wormhole with it. This involves more energy than can be contemplated now, or even in the distant future. Alternately, instead of modifying spacetime as experienced for the rest of the universe, one can merely modify the spacetime around the ship itself to give it the spacetime signature of a tachyon.

The spacetime signature of a tachyon particle is not known at present. It may be a "twist in space" rather than a curved bowl like ordinary matter. In any case converting the starship into a tachyon involves changing the spacetime around the ship from being curved by its own mass to being flat, to finally becoming the tachyonic signature. Since the spaceship weighs only tons, the energy in spacetime curvature is very small, much smaller than the mass energy of a Jupiter-sized planet. In any case, this line of reasoning suggests that FTL travel by intelligent life in this cosmos may be a much less challenging and more widely used technology in the cosmos than previously guessed. If the

Beyond Einstein: GEM Unification

physics of the cosmos allows FTL phenomenon, it is probably already making use of it.

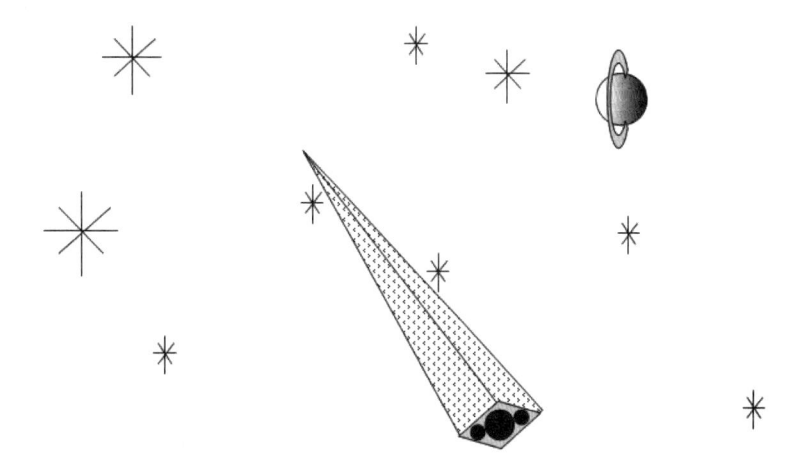

A starship at the edge of the solar system preparing to become tachyonic.

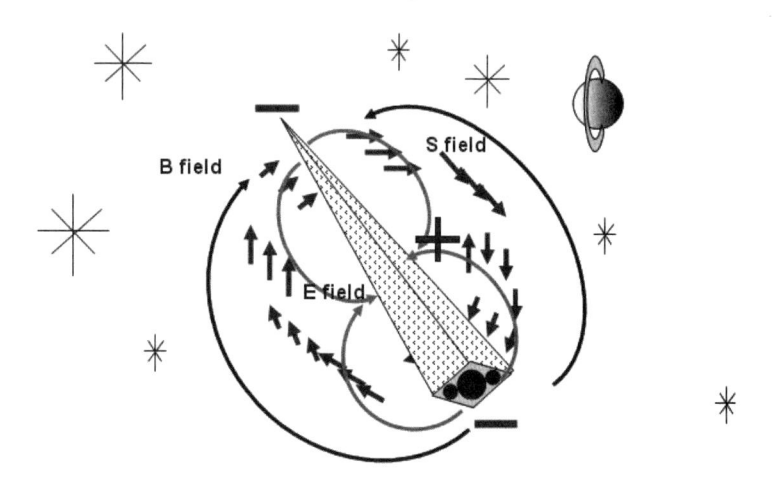

Hypothetical EM fields form around a starship and create spinning Poynting fields as it prepares to become a tachyonic object.

Beyond Einstein: GEM Unification

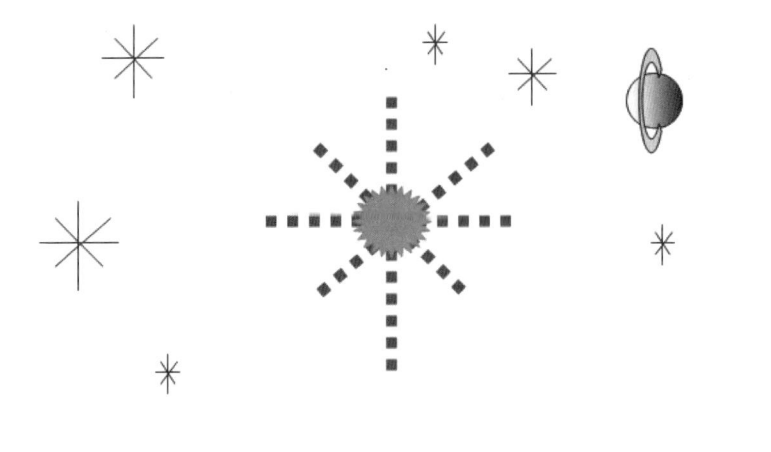

The FTL starship disappears from this spacetime continuum as it goes faster than light.

As it turns out, faster than light travel can probably be done without concerns for causality. Tachyonic particles are probably responsible for the existence of dark energy in the universe, causing it to expand. The expansion of the universe is the clearest indicator of the progress of time. Therefore, to turn one's starship into tachyons to visit the neighbors is to raise the amount of dark energy in the universe, and thus make it age faster. One can, therefore, have breakfast at Neptune, go to Alpha Centauri for lunch, and return to Neptune for dinner, but not breakfast again. The starship may leave the causal stream of the cosmos, but its presence is still felt by the galaxies, which expand more rapidly as it travels. Thus a finite time must elapse between its departure and return, no matter how short.

Beyond Einstein: GEM Unification

Chapter 12. Alpha and Omega: The Cosmos

"Imagination is more important that knowledge."
A. Einstein

"There is no free lunch in the cosmos except the cosmos itself."
Allen Guth

The cosmos is believed to have begun 13 to 14 billion years ago in a tremendous explosion. The kinematics of the heavens, the fact that the galaxies all appear to be retreating from us in an orderly pattern, with a recession velocity increasing with distance, argues that the matter in the universe began in a single point. What we see is similar to what we would see if we were molecules from a piece of high explosive that had exploded in a vacuum and the products of this explosion were now all around us expanding endlessly into emptiness. The array of elements and isotopes in addition to hydrogen that are found in the plasma between the galaxies suggests that the universe was once much more dense and hot, like the inside of a star. Conditions there made, in a split second, helium and lithium as well as heavy

Beyond Einstein: GEM Unification

hydrogen. In one moment, apparently, the cosmos came to be.

What caused this to happen? By what means were we, and all that is, created? Let us start with a pure vacuum. Following the recipe of Sakharov, the universe began with a preference for matter over antimatter, and baryon (heavy particles) and lepton (light particles) quantum numbers were not conserved. The simplest way for this to occur is for the vacuum to split into two charged particles, a plus and a minus, like a matter-antimatter pair. When this splitting occurred, the baryon and lepton numbers began as canceling quantum numbers, like a matter-antimatter pair. Thus the proton and electron were born as a plus and minus version of the same particle and then became different. We will call this particle the union particle. How did this occur?

The universe was born when the fifth dimension was born. The birth of the fifth dimension resulted in a universe full of hydrogen with two long-range forces. When the fifth dimension was born it captured some of the electric and magnetic flux making up spacetime at the Planck scale and expanded it to the subatomic scale. This created the source for gravity and EM fields as we experience them. The fifth dimension deployed from the Planck length to achieve its compacted radius, much, much, larger than the Planck length, but much smaller than the size of the proton or classical electron radius and thus was lost in the quantum haze. The protons and electrons formed and the gravity and EM fields formed as must occur in a five-dimensional spacetime.

However, in 1981, Edward Witten discovered that something occurs when four-dimensional spacetime is increased to become five-dimensional Kaluza-Klein spacetime with a hidden or compact dimension. The new more complex vacuum is now unstable. Witten found that perfectly conducting bubbles of nothingness would appear out of the vacuum and expand. We can predict this based on the idea that, in quantum mechanics, the vacuum is always vibrating and seeking new states, and if a favored length appears, particles with quantum wavelengths in resonance with that length will pop out of the vacuum fueled by the ZPF. Stated differently, the proton and electron are stable particles; however, in a cosmos where EM and gravity are unified, they can occasionally change places and act as each other's antiparticles, causing the vacuum around them to be unstable. The reason for this is again that the quantum vacuum is constantly fluctuating and any hidden dimension or possible particle will not only appear once at the birth of creation, but continue to appear. The presence of a fifth dimension allows the vacuum to have a flaw and decay like a radioactive atom. The result of this very slow decay is that a cubic yard of pure vacuum will emit a spark of light and an electron-proton pair will emerge. The vacuum slowly releases hydrogen.

Edward Witten, the father of vacuum decay.

Beyond Einstein: GEM Unification

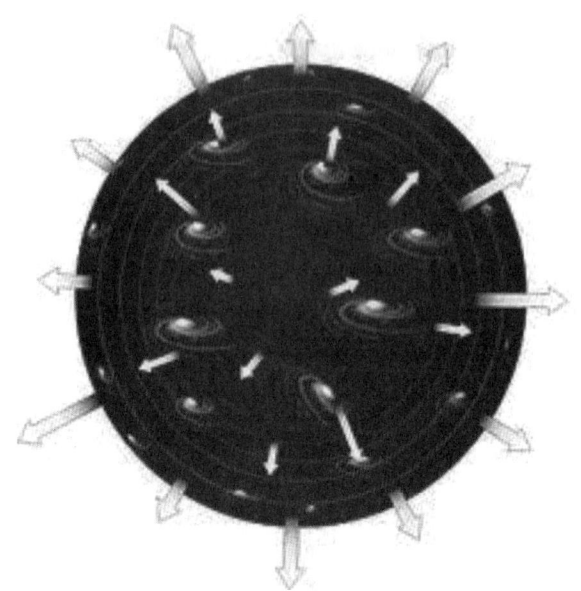

The expanding universe (NASA).

Beyond Einstein: GEM Unification

A.

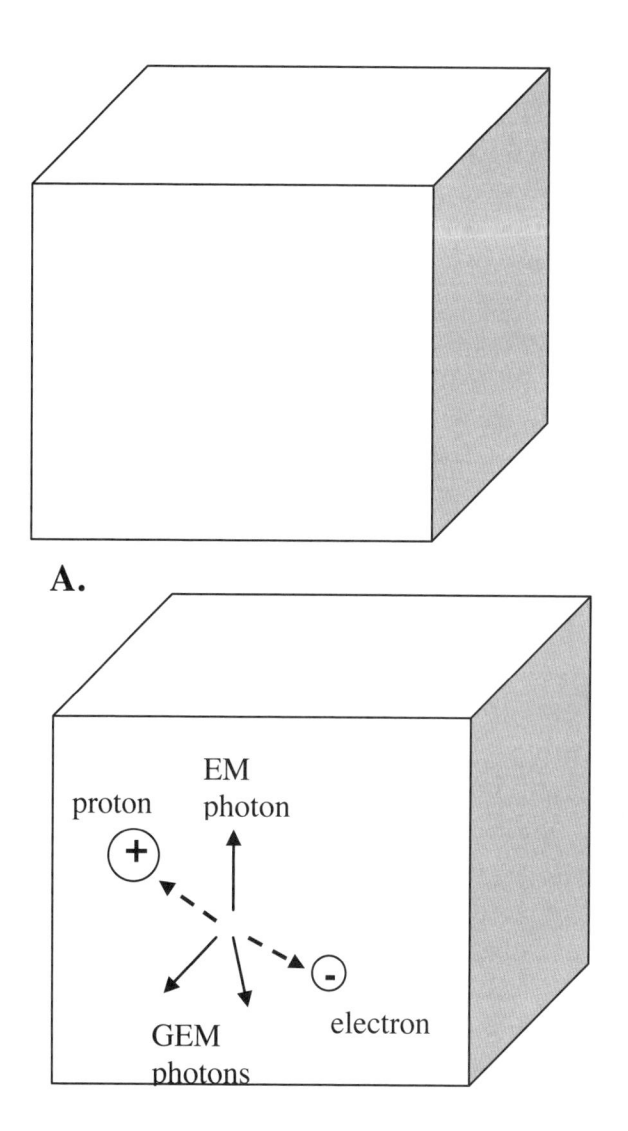

B.

Vacuum decay of a cubic yard of pure vacuum: Once every billion years or so, a spark of light, a gravity ripple, and an electron and proton appear out of the void.

Beyond Einstein: GEM Unification

The moment of creation was different from the present, like the moment of birth of a human being when it ceases to live off the lifeblood of its mother and begins to breathe on its own. Obviously the rate of vacuum decay was much faster then than now. The remnant, in the person, of this primal relationship is the navel or umbilicus. The navel was where the child was connected to its mother, a reminder that the person was once a tiny child. So it is that the universe preserves the hidden dimension, and its length, as a sort of navel. The length of the hidden dimension is still a site of interesting physics and corresponds to particle of rest mass of almost exactly 3000MeV. The Σ (3000) particle, with apparently a very short lifetime, has been seen indirectly, and as an electrically neutral baryon like a neutron, decays ultimately into a proton and an electron. Also nearby are the eta-c meson and the J/ψ meson, which decay either into proton-antiproton pairs or electron-positron pairs.

The vacuum is still unstable, and emits both light and hydrogen periodically. This means on large scales the cosmos is both expanding and gaining new hydrogen to fill the void. If the cosmos is flat and at Ω equals one, then, as pointed out by Hawking, each particle appearing in the universe has zero energy since its gravitational potential energy and mass energy add up to zero. This forms the basis for a universe where new particles can appear without violating conservation of mass-energy. For every unit of mass-energy bound up in electromagnetism when a particle appears in the cosmos from vacuum decay, a negative unit also appears because particles in the universe are in a negative energy state because they are in a gravity well.

Beyond Einstein: GEM Unification

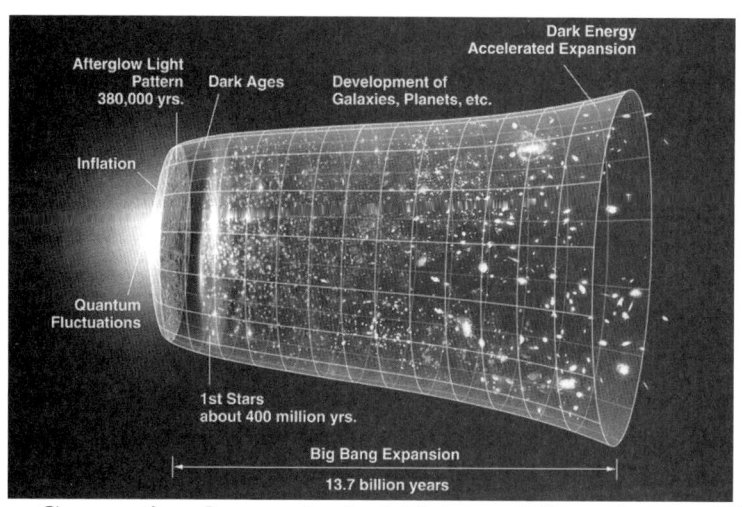

Conventional cosmological history of the universe: The GEM theory merely adds more matter creation as the universe expands, maintaining its density and renewing its supply of hydrogen. (NASA)

When we calculate the rate, according to the GEM theory, that the vacuum will decay to hydrogen, and assume that this maintains an Ω equal to one universe, we find that the Dirac condition results: The ratio of strengths of gravity and EM forces between an electron and proton is the same as the ratio of the size of the Hubble radius, the distance light travels in Hubble time, to the electron. The appearance of the hydrogen as a proton and electron pair means that EM radiation must have been emitted. This emission of EM radiation is controlled by the quantum constant α, which is one over the number 137. In the GEM theory, the resulting Dirac condition relates α and Ω and relates the Planck length to the Hubble radius. The Dirac relation essentially says that the Universe is a creation of gravity just as the electron is a creation of EM and that the two forces are unified.

Beyond Einstein: GEM Unification

With vacuum decay, Hubble time ceases to be an age and becomes a constant of the universe. It was speculated for some time after Dirac's result was published that the Newton gravitation constant or other physical constants were changing as the universe got older. However, Ronald Hellings, using the Viking landers on Mars as beacons, was able to show that no such change occurred with time. This was not only important but ironic, for Ronald Hellings had been the author's inspirational instructor in college many years before, where he served as a visiting professor.

Does all of this mean that the universe has no age? No, not necessarily, it still appears from helium and other light elements that the universe started out much denser and hotter than it is now. So it seems that the universe had to experience a moment of creation. Just as a person breathes now, we also know that every person was a newborn and drew a first breath. So this all means that the universe is now in a steady state expansion, as was first proposed by Hoyle and Bondi. The cosmos is then not ageless, but is effectively eternal; it renews itself like an evergreen tree. It will not end with fire, or ice, it will just keep shining for as long as it is supposed to.

The inflationary hypothesis is also useful to examine. Inflation is a mechanism that allows the universe to expand from the Big Bang with an unstable vacuum and end up at $\Omega=1$. First proposed by Alan Guth, the "Inflationary Scenario" extends the period of non-conservation of mass in the universe from one instant to a tiny fraction of a second. The steady state universe of Hoyle and Bondi, where non-conservation of mass (though not conservation of mass-energy) is continually violated and there is no Big Bang, can then be considered as one end of a continuum of concepts involving the appearance of new particles in the cosmos. This continuum of models, where conventional big

Beyond Einstein: GEM Unification

bang models, have vacuum decay concentrated in one instant of creation. Inflation extends this period of intense vacuum decay over a finite time, and in the Hoyle and Bondi concept make the vacuum decay a continuous, uniform process. It is interesting to note that Lance Williamson of the Konfluence Research Institute has identified an inflationary scenario using GEM concepts, where the cosmos begins as the size of an electron at Compton wavelengths. This would be its natural minimum size once the electron and proton had come into existence.

Allen Guth.

The Cosmic background radiation (CBR) is often cited as evidence for the big bang cosmology, and is indeed important. However, in the context of a vacuum that is five-dimensional and thus unstable, the CBR may represent more than just the afterglow of the big bang. The CBR we see is commonly interpreted as the light from the first split second of the big bang coming from a surface that is beyond the farthest visible galaxies and is receding from us at nearly the

Beyond Einstein: GEM Unification

speed of light, the surface of the Hubble radius. The fact that its temperature is so uniform argues for some mechanism of equilibration over large scales that was operating at FTL (faster than light) since the original universe, under an Einstein or Inflationary model, is expanding much faster than light. It can be argued that in an unstable vacuum scenario, the temperature of the hydrogen appearing from the vacuum would have been set by physical constants such as G and Planck's constant, and the speed of light. However, there is another possible source for the CBR besides the Big Bang afterglow.

The decay of the vacuum that occurred at the big bang is augmented now by the ongoing process of vacuum decay. Each proton and electron pair that pops into being in the cosmos interacts with each other momentarily and emits a photon of EM radiation, a spark of light. If we consider that this spark of light will occur as a completely random, that is to say, thermal process like Hawking decay, then the ensemble of sparks of light will appear over the sky as a thermal EM distribution. We can calculate the thermal radiation being emitted as the pressure of light at the Thompson cross section of an electron as it appears from a miniature expanding universe consisting of one proton and one electron. This calculation shows that the predicted temperature closely matches the observed 2.72 degrees Kelvin. This hypothesis that at least some of the CBR comes from an ongoing process in the cosmos is supported by the observation that galaxies in the line of sight do not block the CBR.

If all the CBR actually came from a surface 13 billion light years away, then a galaxy in the foreground would block most of it. Viewed in CBR, the foreground galaxy would appear as a dark silhouette; however, no silhouettes have been found from distant galaxies. The CBR seems

Beyond Einstein: GEM Unification

unaffected by galaxies in the foreground and is very uniform. This means the source for the CBR lies in the empty space between us and the distant galaxy. The source of some CBR then, is the ongoing process of vacuum decay. When we look anywhere in the sky a long distance, the vacuum along our line of sight is unstable to decay and emission of both hydrogen and the CBR.

This concept of a cosmos that is renewing itself via the vacuum decay process may also explain the dark matter observed in the cosmos. The dark matter is a cloud of particles that seems to exist in the halo around the bright disk of galaxies and adds significant mass to them. The stars orbiting the center of the galaxies do not rotate around the galactic core, as if this was where most of the mass was concentrated. Instead the stars orbit in the disk as if a large amount of mass was located in the halo. The mass distribution must be roughly an inverse square distribution of density around the galactic core, suggesting a compression-less flow inward at uniform velocity. The simplest answer, in the case of a cosmos with an unstable vacuum, is that the dark matter simple hydrogen plasma.

The idea that dark matter may be simply hydrogen in an ionized state explains why it gives no optical signature and also why such matter, which makes up a majority of the mass of the universe, could be comprised of particles not seen in decades of cosmic ray and particle accelerator experiments. The reason, under the vacuum decay into hydrogen hypothesis, is that the particle hides in plain sight; it is the proton.

Beyond Einstein: GEM Unification

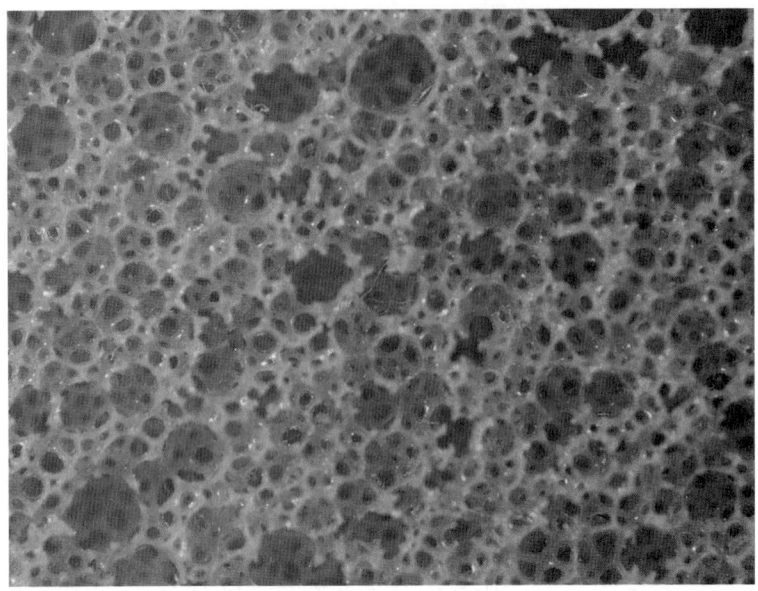

Plastic foam magnified, representative of the EM structure of spacetime near the Planck length in the GEM theory.

The cosmos is of course expanding; in fact, it is exploding. Dark energy, in the form of Einstein's cosmological constant, is the root cause. The cosmos at a large scale looks like a foam, with bubbles of empty space expanding and stretching tendrils of galaxies around them. It is a tribute to Einstein's genius that he was right even when he thought he was wrong. But what is the root cause of the dark energy?

Beyond Einstein: GEM Unification

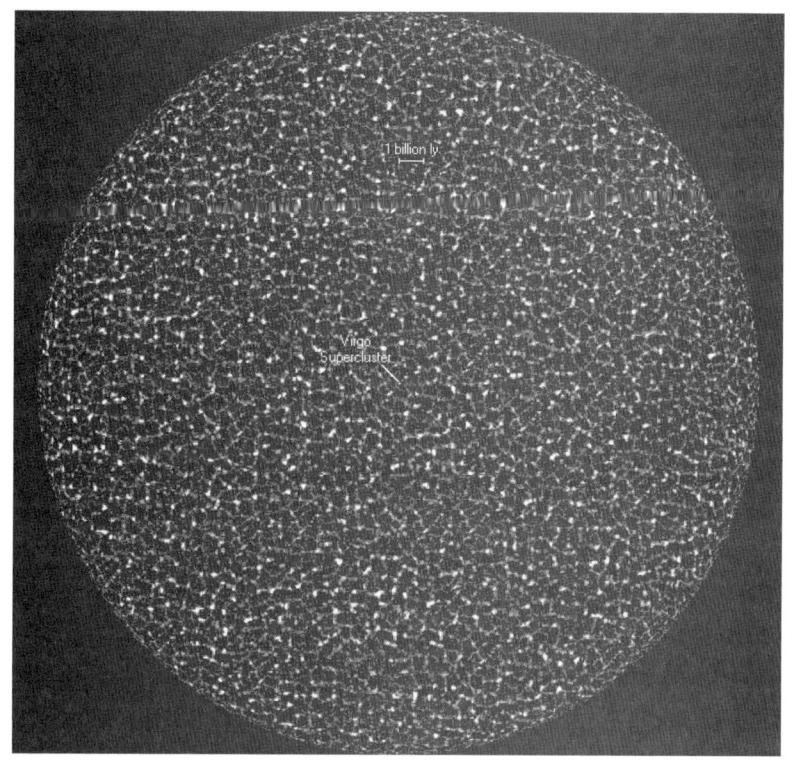

The fibrous structure of the cosmos in large scale is similar to the imagined structure of spacetime itself near the Planck length in the GEM theory.

The simplest mathematical model for the dark energy is that it is due to a vast cloud of tachyons, particles with imaginary mass. These tachyons are mostly at very low energy, and thus transit the visible universe in a heartbeat. These FTL particles are the portion of the cosmos from the big bang that lies beyond what we can see. That particles of imaginary mass would lead to an expansion of the universe can be seen trivially by the fact that the attraction force between two particles is proportional to the product of their masses, and since particle masses are imaginary, the product of their masses must introduce a factor of negative one. The negative one transforms the attraction force of gravity into a force of mutual repulsion. That is, the cosmos is dominated by particles that want to get away from each other as fast as they can. We would see that these imaginary particles, if we could get one to slow down, are probably just hydrogen. The decay of the vacuum is probably just the periodic transition of the tachyons, or, their tunneling into our universe.

Here then is the truth of Einstein's quote, that we live in a cosmos where the imaginary is more important than the real. What is known is real like lights on a Christmas tree on a dark winter night, pretty to look at, but the tree they are hung on is invisible. What drives the cosmos, and the force that lights the lights, is also not visible. The cosmos is alive and full of light, and is renewing continually, and the source of that renewal is imagination.

Beyond Einstein: GEM Unification

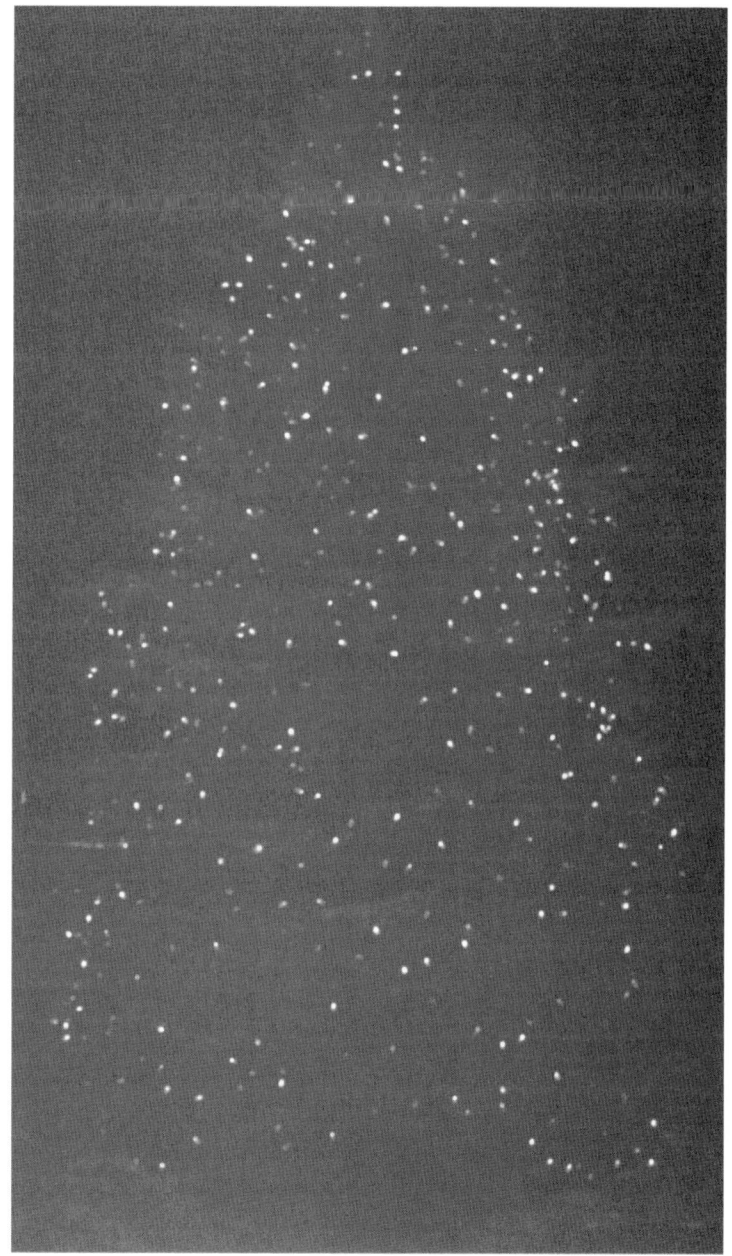

Christmas tree.

Beyond Einstein: GEM Unification

The Great World Tree of the Norse.

Beyond Einstein: GEM Unification

Epilogue: The Summit of Mount Einstein

"The Lord by wisdom has founded the Earth; by understanding hath he established the Heavens."
Proverbs 3:19

Rising early in our tents, we then follow the vale of Sakharov to the Hawking Escarpment, and here find the cleft of Kaluza and Klein, a narrow fissure which allows us a gentle climb. The sky is clear as crystal and the wild mountain winds favorable and gentle this day. . We emerge on the Dirac high glacier, at a great altitude. Oh, what sights we see from even here, including the last slopes to the summit. It is a gentle climb again, and the winds are gentle and do not hamper us as we climb the Sakharov ridge to the summit. As we look back, we see a marvelous sight, for from near the summit we can see that the Sakharov vale, the cleft of Kaluza and Klein, and Sakharov ridge to the summit that had appeared merely as disjointed but happily coincident features for our climb, are, in fact, all one massive structure or rib of the mountain. We decide to call this the Nordstrom structure, after the original discover of the fifth dimension and its wonders. We also see from here the great sheer cliffs of Zeno, and below them the Vortex of Tesla that barred

Beyond Einstein: GEM Unification

Einstein's way though he tried so many times to scale them. They seem the most direct ascent, yet they cannot be scaled, not with any skill we have. There are many paths to a mountain top, but some are more dangerous and difficult than others and we are pleased we have chosen the way we have taken. Yet the summit still belongs to Einstein, for we find to our amazement a sign that he actually reached it, perhaps in the last moments of his life.

Finally we stand at the summit. As we envisioned, the stars are clearly visible in the dark blue sky of the summit, though it be noon day. It is as if the heavens and Earth meet there at the unification summit, and the stars and galaxies seem so close and clear that we could touch them. Truly, in this place that Einstein aspired to, the great long range fields of the cosmos that reach from one end of it to the other are unified, and so is heaven and Earth. We see here that humanity is part of the whole cosmos, not just of Earth. The deep circle of the sky is made one here with the four squares of Earth, never to be divided again.

Einstein, it was said, on his death bed placed his hand over his heart, and said to a friend that he felt he was close to unifying the fields. He spoke in German to his nurse as he breathed his last, but she did not know this language, so his last words are lost.

The GEM theory, however, can be seen as Einstein's final vindication. The Einstein program for unification would not unify only the two fields but unify the electron and proton as well. The GEM theory does this, in a rudimentary way it is admitted, but well enough to allow us to understand gravity fields as composed of EM fields and understand the Newton gravitation constant. This will allow many useful things.

Beyond Einstein: GEM Unification

So we have reached the summit of the mountain. We have unified electromagnetism and gravity, Gravitas, Magnus and Electra, the two long-range forces of nature and with them the two fundamental particles of the universe, protons and electrons. Again, this theory is rudimentary, the author freely admits. It is a "Bohr model" of unification by analogy with the early quantum model of the hydrogen atom. But it is a beginning.

God does not play dice, but He does own the Casino. Dice is played continually in the universe, at the large scale and at the small. Quantum mechanics is as fundamental to physics as Newton's laws, and allows us to understand dynamics far removed from our everyday existence. That Einstein erred when he rejected it is widely believed, though his rejection must be seen as part of his quest for deeper understanding. The GEM theory will not be complete until it is quantized. Gravity alone cannot be quantized; it leads to catastrophes where gravitons, the quanta of gravity, feed on themselves to form black holes. However, EM theory can be quantized, thanks to Feynman, Schwinger, and Tomaga, and has no such problems. Gravity, once unified with EM, will see the black holes evaporate into photons and particle-antiparticle pairs, and should thus be quantizable. In a sense, the Hawking evaporation of the small black hole removes most of the catastrophes of quantum gravity and, once generalized to a Hawking instability of gravitons at short wavelengths, should do the trick. Gravitons will be seen as bundles of EM photons, and photons are well behaved.

The key to understanding the force of gravity in terms of EM is to understand the Poynting force, radiation pressure, which is the third EM force and affects neutral matter. Since the Poynting force is clearly mediated by photons, it becomes, through the quantum mechanics of Feynman, the way to understand all EM forces quantum mechanically, from pure electric to pure magnetic. Because of the insight of Sakharov, we understand that the gravity force is mediated by photons like the EM forces and results from variations of the quantum ZPF. Thus, the quantum Poynting force is seen to underlie the unification of gravity and EM . The fact that the ZPF can be self-censoring and that its variations are perceived as gravity, connects the EM Maxwell tensor to the metric tensor of spacetime. In other words, we understand that gravity, the geometry of spacetime, is unified with EM because spacetime itself is a fabric of electric and magnetic fields. Gravity fields are highly disordered or entropic EM fields, cousins to candle light. EM fields are highly cool, approaching the character of a gravity field only while interacting strongly with large masses of matter, like gravity itself. The unity of the two fields, gravity and EM, can be understood in Feynman's concept that force fields and even the geometry of spacetime is nothing but clouds of photons colliding with charged particles and bouncing off them or being emitted by them. The fields are but clouds of light, some visible, some invisible to us.

The unity of gravity and EM is represented by a star full of hot plasma that sends out EM waves over the whole rainbow of frequencies, as seen by the eyes, but falls off rapidly in the short wavelength, cutting off in the ultraviolet. Likewise the star is held together by gravity, whose rainbow of EM frequencies is hidden in the ZPF and reduces to the Planck length in wavelength. Similar concepts are found in the theory of "loop gravity" championed by Lee Smolin. Hal Puthoff, Bernie Haisch, and Alfonso Rueda, have also made important explorations on ZPF gravity. In general, to satisfy an equivalence principle, the Sakharov concept of gravity as radiation pressure of the ZPF-EM requires that the masses of the particles themselves arise from the ZPF, so

Beyond Einstein: GEM Unification

that the ZPF, in effect, pushes itself. This means that to deeply understand how gravity and EM fields are unified in a complete theory, one must consider the geometric spacetime of particles, not just the fields. This requires a new length to emerge in the cosmos besides the Planck length.

Nikola Tesla.

The GEM theory must consider the structure of particles, to be complete, and the fifth dimension of Kaluza and Klein shows us how this is to be done, for the birth of the fifth dimension gives us not only the two field equations for EM and gravity, but the particles whose dynamics they determine. The birth of the fifth dimension allows the particles and their masses to be born from the spacetime of electric and magnetic fields. They are born capturing electric and magnetic energy from the Planck scale, and separate from both each other and the vacuum. Theoretical Physicist Michio Kaku has argued eloquently for the existence of new dimensions in his excellent book *Hyperspace*. It must be said

Beyond Einstein: GEM Unification

here that the concept of particles as extended objects, not points in a multidimensional spacetime, which is the basic concept of "string or brane" theories attempting to unify gravity, EM, and the strong force, is entirely validated by the concepts of the GEM theory and its numerical results. Therefore, those who ask: *"what has brane theory ever done for physics?"* have their answer: *"it has given you Big G"*

The two particles are born from the vacuum as a vacuum interval of spacetime, with the fifth dimension taking the place of either time or space in the interval. This means one is unitary like time and the other is triune like space. They must have the same charge by Maxwell's equations, but quantum mechanics forbids them to have the same mass. Therefore, one is light and unitary, and is the maker of light itself by its dynamics and the other is heavy and triune, being composed, like space, of three sub-dimensions. Like any three-dimensional object, the proton can be flattened to minimize one or more dimensions, but its three-dimensional character, its volume, is irreducible. So the quarks, which are its sub-dimensions, must remain together. It is like a drop of liquid that is incompressible.

Between the two primordial particles, the electron and proton, the universe has its being. One particle being light, is the source of all light by its lively movements. Its dancing excites ripples of EM waves whenever it moves. This is the electron. The other particle gives all the cosmos inertia and gravity, and is full of entropy and internal dynamics, like the gravity fields it sources. This is the proton. Two particles are born, one time-like, bringing light to the star so it can shine, and the other space-like, a complex heavy particle, giving mass and gravity to the star so it is held together even as it shines. The particles interact in a dance of exchanging photons, some visible, some invisible.

Beyond Einstein: GEM Unification

The new dimension is best understood as charge, being compacted into quanta that can be expressed as a length by using the speed of light and the electron mass. For this reason the electron charge and the charges of the quarks of the triune proton both add up to one as do their squares. The ratio of the size of the new dimension, compact length, one it deploys from the vacuum, is in ratio to the Planck length in its logarithm, 42.8503, and this is mirrored in the square root separation of the proton and electron masses. Thus, the separation of EM from gravity that occurred at the founding of the cosmos can be understood as reflecting in a fun-house mirror, the separation of the electron and proton from the vacuum to form two masses, and from them the whole family of the leptons and baryons. So the deployment of the Kaluza-Klein hidden dimension does double duty, allowing the separation of EM and gravity fields and generating the electron and proton as separate particles. This is reflected in the value of the Newton gravitation constant, which depends on the ratio of the masses of the proton to the electron. However, it also depends on the value of the hidden dimension size to the Planck length. It can be said that the GEM theory has brought the Planck length into the center of physical formula.

The size of the particles and the wavelengths of the EM field associated with light show how far the dimension of particles and their interactions are from the Planck length from which they emerged. This is the mesoscale. To us, the fifth dimension is a hidden, tiny dimension, but to the Planck scale it is enormous, almost astronomical, in size. The success of this program of unifying both the fields and particles in the case of EM and gravity and protons and electrons is a strong endorsement of the idea of hidden dimensions begun by Kaluza and Klein that now continues

Beyond Einstein: GEM Unification

in string theory. But what of the practical import of this unification?

Analysis of the dynamics of particles in a Poynting gravity field reveals that the Kaluza-Klein action can become analogous to a Bernoulli equation and allow the control of gravity fields by Poynting fields. The experiments reported by Kozyrev, Hayasaka, and Takeuchi, and finally, the author, suggest that this equation, like the Bernoulli equation of aerodynamics, can be exploited for flight technology. The obvious shape for craft using this effect for flight near the earth is a flying saucer, which can be considered either a flying satellite dish or a hovercraft using a confined Tesla vortex to generate lifting forces.

The flying saucer shape that emerges raises a number of questions, too numerous to entertain here. The reader is referred to the excellent works of Stanton Freidman for further information. It is rumored that GEM effects may have been discovered during WWII by the Germans, but required too much portable electric power to be useful, and this technology was then captured by the Allied governments. It is also rumored that the Germans may have in turn deduced the basic elements of this technology from wreckage of extraterrestrial craft. Similar rumors exist as to the obtaining of this basic technology by the U.S. government.

Some people think that everyone should know everything, and that the world should be completely transparent, but others think it best that government exercise great discretion in sharing what they know. Loose lips can sink ships. Perhaps the right course lies in between, and humanity evolves to gain new knowledge with some of this knowledge previously held hidden. Truly the cosmos is clear as crystal and the farthest galaxies can be seen, yet they move by dark energies that are invisible. So it is also with the ZPF, which

Beyond Einstein: GEM Unification

cannot become visible without destroying us in the process, and so censors itself to allow us to exist. Yet the ZPF is still in evidence, but in subtle ways. Knowledge is power, but also responsibility. In the sci-fi novel, *Morningstar Pass*, written by the author under the pen name Victor Norgarde, the UFO cover-up collapses, but some of the truths revealed are neither pleasant nor reassuring. It is the nature of truth to emerge and become known, but only the brave can face the full truths about the cosmos we live in. Someday the full story of the flying saucer will emerge, but in the mean time, many engineering challenges await those who can try now to construct one. The circle of the flying saucer has been squared with the laws of physics now known among humanity. It represents a technology within human reach. However, it is one thing to learn the Bernoulli equation, and another to build an airplane that flies, yet another thing to fly it and not crash. When these skills are mastered in the case of the vacuum Bernoulli equation, then the cosmos will become much more accessible to humanity.

It is, as we have noted earlier, amusing and thought provoking that the number that unifies and makes possible a cosmos full of hydrogen with two long-range forces is the number 42.8503... . It is connected to 1/137 from quantum mechanics by a transcendental equation: The Transcendental Cosmos Equation. It says that, in a universe where quantum mechanics exists, the electron and proton can exist. This "transcendental" number is approximately the number 42 said to be the answer to the "great question" in the highly entertaining series of sci-fi novels, *The Hitchhikers Guide to the Galaxy*. Thus, in this book at least, the question and its answer are connected on one page: *"what number allows a quantized universe full of hydrogen with two long-range forces to appear out of the vacuum, where Gravitas, Magnus and Electra, frolic amid the waters above and beneath, so*

Beyond Einstein: GEM Unification

that this great cosmic epic can occur for God's purposes?" The answer is *42.8503… .*

Albert Einstein.

Beyond Einstein: GEM Unification

The image of the cosmos that emerges after GEM unification is a rich and exotic one. The cosmos apparently began as a vacuum with only gravity, yet within gravity EM fields were concealed; then, with the advent of the fifth dimension came a vast and violent explosion when both hydrogen and EM fields filled the void. It now continues to expand from the force of this explosion. However, the very appearance of the fifth dimension that triggered the vast explosion has made four-dimensional spacetime unstable so that hydrogen even now pops out of the vacuum as the universe expands. When the rates are calculated approximately, the steady state universe of Hoyle is recovered with the Dirac relation as a signpost that EM and gravity forces are conspiring together to run it. The Dirac relation means that the particles of the cosmos must be critically connected to each other by EM in order for them to be critically connected by gravity. Similarly, the Dirac relation is the outward sign of the Witten instability introduced by the existence of the fifth dimension into the cosmos that allows the vacuum to decay into hydrogen as it expands, renewing itself in the process. The CBR is thus the glow of new matter flowing into the universe from some hidden reservoir, rather than only the afterglow of the big bang. For this reason, its sources are in the space near to us, not only from the far edge of the universe. The Hubble time now serves as a dynamic constant rather than just a time from the beginning. The cosmos is like a lit candle that now burns continually and renews itself. The cosmos will not die the death of fire or ice, it will instead burn brightly for the time appointed to it.

The dark energy that powers the universe's expansion is due to FTL particles that form the reservoir for the matter that appeared. So it is that FTL phenomenon is not just

Beyond Einstein: GEM Unification

possible for humanity to travel to the stars, it is the lifeblood of the cosmos.

Much more could be said about this. A whole library can be written and will be. But for now the story of the finding of this thing, the unifying of gravity and EM, as Einstein said was possible, has been done. Einstein like any human had his failings, but he was not afraid to fail. Einstein was right, not in detail, but in the whole story, for the unification was made possible by the very elements that he had made available to the physics community: relativity, the ZPF, and the Kaluza-Klein theory, and, finally, the intense conviction and utterance that such a unification was possible. Thus is the tale told: Einstein was right all along, perhaps not in some details, but in his overarching vision, he was right.

As we descend from the summit, the weather remains good. What sights we have seen on our journey, from the Planck length to the Hubble radius! But one of our party has observed the GEM's unification peak in the distance, and he continually harangues us that we must try to summit this peak next year. We are weary, but his speeches as we descend are finally heartening.

"Is not the Compton wavelength of the pion, the carrier of the strong force, the classical radius of the electron?" he yells over the winds.

"Does not the neutral pion decay into pure light? Is not the mass ratio of the proton to the electron derivable to close accuracy from a Planckian photon field in the proton?"

"Does not the sum of the charges of the quarks add to one, but also their squares?" another calls out.

"Surely this says that the strong force can be unified with Gravo-Electro-Magnetism! Why, it fairly cries out that they should be unified! What Ho! From this height I see a path to the summit of the GEMs mountain in the distance, through

Beyond Einstein: GEM Unification

the passes of Gell-Mann, and the glaciers of Yang-Mills, up the cliffs of the Popov Ghosts, and via the T'Hooft process to the pinnacle!"

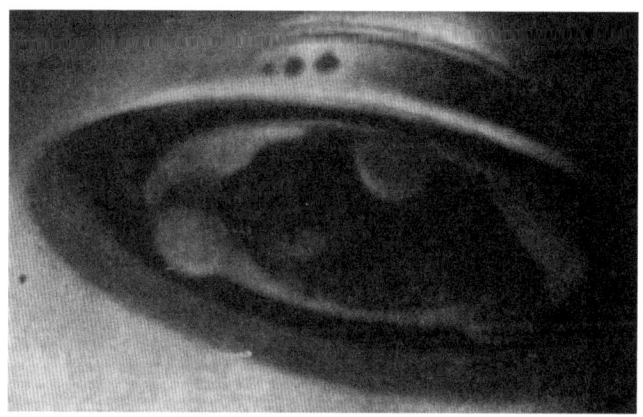

UFO: Any questions?

Beyond Einstein: GEM Unification

Beyond Einstein: GEM Unification

Afterword: Seven Numerical results of the GEM theory

1. Relation of the proton-electron mass ratio to the 5th dimension size

$$r_0: \left(\frac{m_p}{m_e}\right)^{1/2} = \ln\left(r_o/r_P\right) = 42.8503...$$

2. The value of the Newton Gravitation Constant (to leading order in cgs units)

$$G = \left(e^2/(m_p m_e)\right) \alpha \exp\left(-2\left(m_p/m_e\right)^{1/2}\right) = 6.668 \times 10^{-8}\, dyne-cm^2 g^{-2}$$

3. The Transcendental Cosmos equation (see appendix for derivation)

$$\sigma \cong \ln \sigma(\alpha^{-1/2} + 1) - \ln \alpha^{-1}$$

4. The Self-Censorship relationship of the ZPF EM stress tensor to the spacetime metric

$$g_{\alpha\beta} = \frac{4 F_\alpha^\gamma F_{\gamma\beta}}{F_{\mu\gamma} F^{\mu\gamma}}$$

Beyond Einstein: GEM Unification

5. The Vacuum Bernoulli Equation

$$\frac{g^2}{2\pi G} - \frac{S^2}{u_o c^2} = Cons\tan t$$

6. The Hubble Time

$$T_H = \left(\frac{9}{8\pi}\left[\frac{1}{\sigma\alpha}\right]^4 \frac{1}{\alpha}\right)^{1/3} \frac{e^2}{Gm_e m_p} \frac{r_e}{c} \cong 12 \times 10^9 \, yrs$$

7. The CBR temperature

$$T_{CBR} = \left(\frac{3}{4}c\left[\frac{Gm_e^2}{\sigma_{Th}^2 \sigma_{Stef-Boltz}}\right]\right)^{1/4}\left[1+\frac{1}{\sigma}\right] = 2.72K$$

Beyond Einstein: GEM Unification

Chapter notes

Prologue

The volume of the gold torus is $6\pi^5 = 1836.12$ units which is within 17 parts per million of the actual value of the measured proton-electron mass ratio 1836.15, This formula was first proposed by Lenz, a German Physicist, in 1951 its square root is $\sigma = 42.8503\ldots$

Chapter 1

The Cosmos gives clues as to its origins and fundamental processes. The Cosmos is expanding with a "Hubble flow" carrying the galaxies apart from each other with a velocity V described by the equation

$$V = \frac{R}{T_{Hubble}} \quad (1)$$

where R is the distance between the galaxies and T_{Hubble} is the Hubble time. Recently, it has been discovered that the rate of expansion is increasing believed driven by "Dark Energy." At the same time, the Cosmos appears approximately uniform in density, over large scales, is close to the condition of critical density such that if one defines a Hubble radius $R_H = cT_{Hubble}$, you can define a sphere about any point a quantity Ω

$$\Omega = \frac{8\pi}{3}\frac{Gnm_p}{c^2}R_H{}^2 = \frac{V^2}{c^2} \qquad (2)$$

Where n is the cosmic density of a hydrogen plasma

We can also write a parameter for the degree of EM interaction within the cosmos at low energy, which is the probability that an electron will emit a low energy photon and have it scatter off another electron. If we say the parameter Ω is equal and correlated to the EM interaction parameter (the Thomson scattering parameter) then we have

$$\Omega \cong \sigma_{Th} n R_H \qquad (3)$$

Where σ_{Th} is the Thomson cross section of the electron, and n is again the density of hydrogen plasma in the cosmos. If one requires that the Cosmos degree of EM self-interaction be correlated with its degree of gravitational self-interaction, that is: exchanging momentum and energy via photon emission and absorption, one recovers the relationship first discovered by Dirac, called the LNH (Large Number Hypothesis.)

$$\Omega = \frac{8\pi}{3}\frac{Gnm_p}{c^2}R_H{}^2 \approx \sigma_{th} n R_H$$

(3)

$$r_o = \frac{e^2}{m_e c^2} \qquad (4)$$

$$\frac{Gm_p m_e}{e^2} \approx \frac{r_o}{R_H} \tag{5}$$

Thus the meaning of the Dirac Large Number Hypothesis is that the critical self-binding due to gravity in the Cosmos is correlated to the fact that it is critically EM self-interacting. This suggests gravity is fundamentally connected to EM interactions. The appearance of the composite mass $m_o^2 = m_p m_e$ suggests that this mass is also fundamental to the dynamics of the Cosmos. From these two ideas we can derive the GEM theory.

Chapter 2

The figure "Behold" proves the Pythagorean theorem in one diagram.

TOTAL AREA = $(a+b)^2 = a^2 + 2ab + b^2$ = $a^2 + b^2 + 4(ab/2) = c^2 + 4(ab/2)$, $4(ab/2)$ = area of triangles

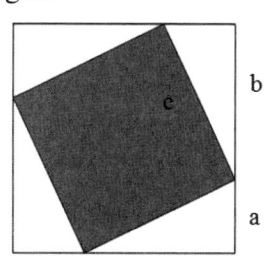

Beyond Einstein: GEM Unification

To understand the relationship of gravity (curved space) to EM (rectified space) is to complete the poetic truth of the letter G, which is the truth of the squared circle.

G →

G → G → ⬚

Chapter 5

$$\alpha = \frac{e^2}{4\pi\varepsilon_o \hbar c} \cong \frac{1}{137}$$

Chapter 7
If we equate the Compton wavelength of a massive particle to its Schwartzchild radius we obtain, the Planck length and the Planck Mass:

Beyond Einstein: GEM Unification

$$\frac{GM_P}{c^2} = r_P = \frac{\hbar}{M_P c}$$

$$r_P = \sqrt{\frac{G\hbar}{c^3}} \qquad M_P = \sqrt{\frac{\hbar c}{G}}$$

Chapter 10

The fundamental relation of the GEM theory is the relationship between the square root of the proton-electron mass ratio: $m_o = (m_p/m_e)^{1/2}$ and the ratio of the size of the fifth dimension to the Planck length r_o/r_P.

$$r_P = \sqrt{\frac{G\hbar}{c^3}}$$

$$r_o = \frac{e^2}{4\pi\varepsilon_o m_o c^2}$$

$$\left(\frac{m_p}{m_e}\right)^{1/2} = \ln\left(r_o/r_P\right) = 42.8503...$$

The ratio of the 5th dimension size to the Planck length is also a quantum-normalized ratio of the gravity and EM coupling constants, where α is the fine structure

293

constant and m_o is the union mass: $m_o = (m_p m_e)^{1/2}$

$$r_o/r_P = \sqrt{\frac{e^2 \alpha}{G m_o^2}}$$

We can obtain from this relationship an expression for the Newton Gravitation Constant

$$G = \left(\frac{e^2}{4\pi\varepsilon_o m_p m_e}\right) \alpha \exp\left(-2\left(\frac{m_p}{m_e}\right)^{1/2}\right) = 6.668 \times 10^{-11} Newton - m^2 kg^{-2}.$$

This formula, in MKS units, gives G to within 1 part per thousand of its presently accepted value of 6.674×10^{-11} Newton-m^2 – kg^{-2}, it was first presented at an APS meeting in 1988.

Using esu units:

$$e^2/4\pi\varepsilon_o \,(MKS) = e^2 \,(esu)$$

$$G = \left(\frac{e^2}{m_p m_e}\right) \alpha \exp\left(-2\left(\frac{m_p}{m_e}\right)^{1/2}\right) = 6.668 \times 10^{-11} Newton - m^2 kg^{-2}$$

This means the gravity force between two masses M_1 and M_2 can be modeled as being due to a modified electrostatic force, where each mass is divided my a standard mass m_o

Beyond Einstein: GEM Unification

$= (m_p m_e)^{1/2}$ and then assigned an imaginary charge e with a reduced probability of photon exchange P_{GEM} to account for the weakness of gravity between small masses.

$$P_{GEM} = \frac{Gm_p m_e}{e^2} = \alpha \exp\left(-2\left(m_p/m_e\right)^{1/2}\right)$$

$$Force = -\frac{GM_1 M_2}{R^2} = \frac{e^2 (iM_1/m_o)(iM_2/m_o)}{R^2} P_{GEM}$$

Note that the P_{GEM} is linear in α, the probability of an EM photon exchange indicating that gravity is a higher order quantum EM interaction, as was proposed by Sakharov.

Self-Censorship of the ZPF

The EM stress tensor for the ZPF is zero to first order:

$$T_{\alpha\beta} = F_\alpha^\gamma F_{\gamma\beta} - g_{\alpha\beta} \frac{F_{\mu\gamma} F^{\mu\gamma}}{4} = 0$$

In the GEM theory this occurs because part of the ZPF stress tensor becomes the spacetime metric

Beyond Einstein: GEM Unification

$$g_{\alpha\beta} = \frac{4F_\alpha^\gamma F_{\gamma\beta}}{F_{\mu\gamma}F^{\mu\gamma}}$$

In this way, an ultra-strong field self-censors itself by becoming the ground of reality

The Planck Charge q_v can be defined and its ratio to the electron charge is

$$\alpha^{-1/2} = \left(\frac{\hbar c}{e^2}\right)^{1/2} = \frac{q_v}{e}$$

The mass of the proton, which carries nearly all the mass of the electron-proton system, should be logarithmically related to the Planck mass by the ratio of the "vacuum charge", to the particle scale charge, to leading order in $1/\sigma$:

$$\ln\left(\frac{M_P}{m_p}\right) \cong \alpha^{-1/2} \ln \sigma$$

This leads then with the equation for 5th dimension size to the ***Transcendental Cosmic Equation:***

$$\sigma \cong \ln\sigma(\alpha^{-1/2}+1) - \ln\alpha^{-1}$$

This occurs because of relationship of Planck scale mass and length to mesoscale quantities (to leading order). See appendix article for derivation.

Which determines the number $\sigma = 42.85...$ from the value of $\alpha \sim 1/137$, and which determines the creation of the mesoscale quantities, e, m_p, and m_e form the Planck scale. The equation is shown here only to leading order in σ, the more precise version

Beyond Einstein: GEM Unification

is shown in the following extended GEM article.

We consider r_o is a Compton radius and look for the particle mass that corresponds to it:

$$\frac{\hbar}{mc} = r_o$$

$$mc^2 = \frac{\hbar c}{r_o} = \frac{\hbar c \sqrt{m_e m_p} c^2}{e^2} =$$

$$\frac{1}{\alpha}\left(\frac{m_p}{m_e}\right)^{1/2} m_e c^2 = 3000.6 MeV$$

Chapter 11

The Vacuum Bernoulii Equation is derived most simply from the Kaluza-Klien Action, which fro weak field s can be reduced

$$(16\pi G)^{-1} K_{field} = \frac{R}{16\pi G} + \frac{E^2 - B^2}{8\pi} = \frac{g^2}{2\pi G} - \frac{S^2}{u_o c^2} = C$$

where g is the local gravity, S is the Poynting vector and u_o is the local magnetic field energy density, and C is a constant. This can be contrasted with the aerodynamic Bernoulli Equation

$$P + \frac{1}{2}\rho v^2 = C$$

Note the negative sign

Beyond Einstein: GEM Unification

Chapter 12

To relate α to Ω, that is the Planck scale to Cosmic scale, we begin with the Dirac Large Numbers expression for the Hubble Time

$$\frac{e^2}{Gm_e m_p} \frac{r_e}{c} = 2.3x10^{39} x \frac{2.82x10^{-15} m}{3x10^8 m/\sec} = 2.3x10^{39} x 0.94x10^{-23} = 2.16x10^{16} \sec$$

$$1 yr = 3.16x10^7 \sec$$

$$\frac{e^2}{Gm_e m_p} \frac{r_e}{c} = \frac{2.16x10^{16} \sec}{3.16x10^7 \sec/yr} = 6.8x10^8 \, yr = 0.68x10^9 \, yr$$

we now assume for some value of Ω ~ 1

a steady state density inflow to the cosmos from the vacuum

$$\frac{8\pi G m_p n R_H^2}{3} = c^2 \Omega \qquad n_c = \frac{3c^2 \Omega}{8\pi G m_p R_H^2}$$

$$\frac{1}{T_H} = \frac{c}{R_H}$$

$$n_{,t} = \frac{3n_c}{T_H} = \frac{9c^2}{8\pi G m_p R_H^2 T_H} = \frac{9\Omega}{8\pi G m_p T_H^3}$$

We now assume that in each Compton oscillation in a Compton volume there is a probability P_{GEM}^2 that a GEM photon will be scattered off a union particle turning it into a proton and another will scatter off a union turning it into an electron and that the resulting charged particles will interact through their EM fields. This rate is also controlled by Ω, which is the self-interaction probability of the cosmos. This means a density inflow from the ZPF at a rate

Beyond Einstein: GEM Unification

$$n_t = \frac{\Omega c}{\lambda_o^4} P_{GEM}^2 \alpha = c\Omega \left[\frac{\sigma\alpha}{r_e}\right]^4 P_{GEM}^2 \alpha$$

We equate these two expressions for density inflow from the vacuum, one to maintain Ω in terms of gravity, and the other a quantum fluctuation controlled by Ω, and we obtain for a steady state cosmos a Hubble Time T_H expression:

$$T_H^3 = \frac{9}{8\pi}\left[\frac{1}{\sigma\alpha}\right]^4 \frac{1}{\alpha}\left[\frac{e^2}{Gm_e m_p}\right]^3 \frac{r_e^3}{c^3}$$

Or in final form

$$T_H = \left(\frac{9}{8\pi}\left[\frac{1}{\sigma\alpha}\right]^4 \frac{1}{\alpha}\right)^{1/3} \frac{e^2}{Gm_e m_p} \frac{r_e}{c} \cong 12 \times 10^9 \text{ yrs}$$

The expression for the CBR temperature is derived in both first and second GEM articles (1991, 1996) to zeroth order, a first order correction in $1/\sigma$ is added here.

$$T_{CBR} = \left(\frac{3}{4}c\left[\frac{Gm_e^2}{\sigma_{Th}^2 \sigma_{Stef-Boltz}}\right]\right)^{1/4}\left[1 + \frac{1}{\sigma}\right] = 2.72K$$

Beyond Einstein: GEM Unification

Epilogue

Basically, The Lenz expression for the proton-electron mass ratio hides within the Stefan-Boltzmann constant of Planck's spectrum. But this is the next adventure.

$$\sigma_{Stef-Boltz} = \frac{\pi^2 k_B^4}{60\hbar^3 c^2} = \frac{(6\pi^5) k_B^4}{45 h^3 c^2}$$

The quark charge system of the proton

$$q_x + q_y + q_z = 1$$
$$-1/3 + 2/3 + 2/3 = 1$$
$$q_x^2 + q_y^2 + q_z^2 = 1$$
$$1/9 + 4/9 + 4/9 = 1$$

Chapter bibliography

Prologue
Denis Brian, 1996, *Einstein, A Life*, John Wiley and Sons, New York, New York.
Lenz, F., 1951, Phys. Rev., v82, p554. The proton-electron mass ratio

Chapter 1
Carl Sagan, Cosmos
J.E. Brandenburg (1992) "Unification of Gravity and Electromagnetism in the Plasma Universe" IEEE Transactions on Plasma Science, Vol 20, 6, p944.

Chapter 2
Cosmos

Chapter 3
Cosmos

Chapter 5
Robert P. Crease and Charles C. Mann, (1986) *The Second Creation*, MacMillan Publishing Co. New York, NY

Chapter 6
Sakharov, A.D., 1967, Doklady Akademii Nauk SSSR, v177, n1, p70.
Vacuum Quantum Fluctuations in Curved Space and the Theory of Gravitation

Chapter 9
Hans, C. Ohnian, (2008) *Einstein's Mistakes: The Human Failings of Genius*

W. W. Norton & Company; NY NY.

Chapter 10
A THEORETICAL VALUE FOR THE NEWTON GRAVITATION CONSTANT FROM THE GEM THEORY OF FIELD UNIFICATION AND THE KURSUNOGLU-BRANDENBURG HYPOTHESIS OF MASSIVE GAMMA-RAY BURSTERS THE

Beyond Einstein: GEM Unification

LAUNCHING OF LA BELLE EPOQUE OF HIGH ENERGY PHYSICS AND COSMOLOGY, A Festschrift for Paul Frampton in His 60th Year and Memorial Tributes to Behram Kursunoglu (1922-2003) (pp 112-119) J. E. BRANDENBURG

A Second-Order-of-Accuracy Derivation of the Newton Gravitation Constant From The GEM Unification Theory, John Brandenburg Bulletin of the American Physical Society Spring 2010 Meeting of the Ohio Section of the APS Volume 55, Number 4

"The value of the Newton Gravitation Constant and its Relationship to Cosmic Electrodynamics", Brandenburg,J.E. (2007) IEEE Trans Plasma Sci, Vol. 35, No. 4., p845.

Clark M.J. " A Theroretical Formula for the Gravitation Constant Based on a Boson Theory" from The Gravitation Constant: Porceedings of a NATO Summer Study, Granazzli H.and Gilley A. Editors (2003) **Chapter 11**

Beyond Einstein: GEM Unification

Faster Than Light (FTL) Travel and Causality in the Context of the Gravity-Electro-Magnetism (GEM) Theory of Field Unification , Brandenburg, J. E.SPACE, PROPULSION & ENERGY SCIENCES INTERNATIONAL FORMUM SPESIF-2010: 14th Conference on Thermophysics Applications in Microgravity 7th Symposium on New Frontiers in Space Propulsion Sciences 2nd Symposium on Astrosociology 1st Symposium on High Frequency Gravitational Waves. AIP Conference Proceedings, Volume 1208, pp. 350-358 (2010).

Brandenburg , J.E. , "The GEM Theory of Unification of Gravity and Electromagnetism" 2010 meeting of the Division of Particles and Fields of the APS Wayne State University

Chapter 12

J.E. Brandenburg (1995) " A Model Cosmology Based on Gravity Electro-Magnetism Unification", Astrophysics and Space Science, Vol 277, p133-144. (also in "Plasma Astrophysics and Cosmology" (1993)A. L. Peratt Editor , Kluwer Academic Publishers)

The GEM (Gravity-Electro-Magnetism) Unification Theory and Cosmic Baryo-Genesis Brandenburg Bulletin of the American Physical Society Spring 2010 Meeting of the Ohio Section of the APS Volume 55, Number 4

Beyond Einstein: GEM Unification

R. W. Hellings, P. J. Adams, J. D. Anderson, M. S. Keesey, E. L. Lau, and E. M. Standish

Phys. Rev. Lett. 51, 1609–1612 (1983)
Experimental Test of the Variability of G
Using Viking Lander Ranging Data
Epilogue
Robert P. Crease and Charles C. Mann,
(1986) ***The Second Creation***, MacMillan
Publishing Co. New York, NY

A Theoretical Value for the Newton
Gravitation Constant from the GEM
Unification Theory of Gravity and Electro-
Magnetism, Brandenburg J.E. OSS09 2009
Meeting of the Ohio Chapter of the
The American Physical Society

Beyond Einstein: GEM Unification

A Theory of the Unification of Gravitation and Electromagnetism

J.E. Brandenburg

Orbital Technologies Incorporated
Madison, WI, 53717, USA

Abstract: The GEM (Gravity Electro-Magnetism), theory is presented as an alloy of Sakharov and Kaluza Klein approaches to field unification. GEM uses the concept of gravity fields as Poynting fields to postulate that the non-metric portion of the EM stress tensor becomes the metric tensor in strong fields leading to "self-censorship". Covariant formulation of the GEM theory is accomplished through definition of the spacetime metric tensor as a portion of the EM stress tensor normalized by its own trace: $g_{ab} = 4(F^c{}_a F_{cb})/(F^{ab}F_{ab})$, it is found that this results in a massless ground state vacuum and a Newtonian gravitation potential $\phi = \frac{1}{2} E^2/B^2$. It is found that a Lorentz of flat-space metric is recovered in the limit of a full spectrum ZPF (Zero Point Fluctuation). The vacuum ZPF energy and vacuum quantities G, $\hbar$, c gives birth to particles of quantities m_p, m_e, e, and $-e$ in a process triggered by the appearance of the Kaluza-Klein fifth dimension. Where also the EM and gravity forces split from each other in a process correlated to the splitting apart of protons and electrons. The separate appearance of the proton and electron occurs as the splitting of a light-like spacetime interval of zero length into a finite space-like portion containing three subdimensions identified with quarks and a time-like portion identified with the electron. The separation of mass with charge for the electron and proton pair comes about from a U(1) symmetry with a rotation in imaginary angle. A logarithmic variation of charge with mass for the proton-electron pair results and leads to the formula $\ln(r_o/r_p) = \sigma$, where $\sigma = (m_p/m_e)^{1/2}$, where m_p and m_e are the electron and proton masses respectively and where $r_o = e^2/m_o c^2$, and where $m_o = (m_p m_e)^{1/2}$ and the formula : $\ln(M_P/m_o) \cong \ln\sigma(\alpha^{-1/2} + 1)$. GEM theory is now validated through the Standard Model of physics. Derivation of the value of the Gravitation constant based on the observed variation of α with energy results in the formula $G \cong \hbar c/M_{\eta c}^2 \exp(-1/(1.61\alpha))$ where α is the fine structure constant; $\hbar$ is Planck's constant; c is the speed of light; and $M_{\eta c}$ is the mass of the η_{cc} Charmonium meson and is shown to be identical to that derived from GEM postulates. GEM is consistent with quantum renormalization with an ultraviolet cutoff at the Planck length. Perturbation theory leads to the formula $G = e^2/m_o^2 \alpha \exp(-2\sigma)(1+0.86/\sigma^2) = 6.674 \times 10^{-11}$ dynes-cm^2/g^2 which is within experimental accuracy.

Beyond Einstein: GEM Unification

I. Introduction: The Postulates of the GEM Theory

"Depend upon it, there is nothing so unnatural as the commonplace."
Sherlock Holmes, in "A Case of Identity"

The commonplace things of the cosmos—hydrogen, EM, and gravity fields—because of their great stability and conservative nature, present us with a paradox, for their very stable and ubiquitous nature require an equally violent and massive process to have brought them into being. In this article I will attempt to explain how the proton and electron, and the long range fields between them, came into being. The GEM theory (Brandenburg 2007, 1996, 1991)is a geometric theory that is an alloy of the Sakharov (1967) and Kaluza-Klein (1929) approaches to the unification of EM and gravity, the two long range forces of nature. The theory is fairly primitive, being described as a "Bohr Model" of field unification, by analogy to the early simple model of the quantum mechanics of the hydrogen atom". It is not "a theory of everything" but rather "a theory of most of what is observed. The GEM theory is not just a theory of unification of the two long range forces of nature, but of scales, for GEM also stands also for "grandis et medianis," the unity of the great and the middle, because it attempts to relate the meaning of Planck and cosmic scales with the middle or mesoscale of classical particle radii. In this theory the most commonplace things of the cosmos suggest processes that are the most far removed from natural experience.

The GEM theory, its full covariant form, is seen properly as a theory of the origin of both the most common and stable forms of matter , protons and electrons, and long range force fields, gravity and EM, from the vacuum. The fact that, in GEM, the vacuum itself is unstable to the appearance of

Beyond Einstein: GEM Unification

particles, leads to a present state of inflation, and cannot be divorced from questions of the evolution of the cosmos, as has been previously discussed (Brandenburg 1998)

The two postulates of GEM in its original form (Brandenburg 1991): 1, Gravity consists of arrays of varying ExB drift cells or Poynting fields such that ultra-strong fields will self censor; 2, both EM and gravity fields and electron and proton quantum fields unify at the Planck scale into one field and one particle. They split apart to form two distinct fields and two distinct particles with the appearance of a new fifth dimension and have been found to lead respectively to two corresponding expressions: 1, The VBE (Vacuum Bernoulli Equation) that predicts EM control of spacetime structure (Brandenburg and Kline 1998); and 2, A Gravitation Constant formula (Brandenburg 1991).

In the new, covariant form, of GEM, it is found that the concept of Sakharov (1967) that the "metric elasticity of space" is due to the EM energy of ZPF (Zero-Point Fluctuation) spacetime leads to the revised postulate 1, the metric tensor in 4 space is a normalized form of the ZPF EM stress tensor such that strong fields "self censor"; and for the second postulate 2, the appearance of the Kaluza-Klein fifth dimension comes from the splitting of a compact "light-like" spacetime interval, the only spacetime interval compatible with the vacuum, and that this triggers the appearance of both the proton and electron from the vacuum ZPF and the separate appearance of the EM and gravity force.

GEM Theory Overview

We begin with the quantities that characterize the vacuum: the gravitation constant G, the rationalized Planck's constant, $\hbar$, and the speed of light c, which can be combined to yield the vacuum quantities of the Planck length, the Planck Mass and a "vacuum charge" respectively:

Beyond Einstein: GEM Unification

$$r_P = \sqrt{\frac{G\hbar}{c^3}} \quad (1a)$$

$$M_P = \sqrt{\frac{\hbar c}{G}} \quad (1b)$$

$$q_v = \sqrt{\hbar c} \quad (1c)$$

At The Planck scale in vacuum, quantization of even the geometry of spacetime itself is accomplished, time and space are quantized in r_P, r_P/c, and mass in M_P.

The vacuum can be described as a region free of matter and the fields that arise from it. The vacuum is then disturbed by the appearance of matter in the form of particles. In the GEM theory we make the physical definition of a stable particle as a region of space smaller than other physically relevant lengths where physical properties occur that do not occur in the surrounding space. These properties or physical quanta, are independent of the particle movement in time t, or space, x, y, z coordinates and can thus be described as a fifth dimension orthogonal to the other coordinates. We restrict our theory to two stable particles, the electron and proton, that together make up hydrogen and can be viewed as the building blocks of 99% of the observed matter in the cosmos.

The size of the particle, or 'hidden" dimension, is a constrained spacetime interval scale where the fifth dimension can mix freely with the normal 4 dimensions of spacetime. It is more of a small doorway than a hidden dimension. This can be described in two limits as a quantized radius of a small sphere r_o or a time interval t_o.

$$r_o^2 \leq (x_o^2 + y_o^2 + z_o^2) \quad (2a)$$

$$r_o^2 \leq c^2 t_o^2 \quad (2b)$$

Beyond Einstein: GEM Unification

The GEM theory will describe how the full Maxwell's and general relativity equations and the non-vacuum quantities of the electron and proton charge –e, e, and the electron and proton mass m_e and m_p arise with the birth of the Kaluza - Klein fifth dimension , which defines what are called particles. These give rise to a new length scale, r_o which we will term the "mesoscale" because it is intermediate between the Planck and Cosmic scale and which consider to be the hidden dimension size and a new mass scale m_o

$$m_o = \sqrt{m_p m_e} \qquad (3a)$$

$$r_o = \frac{e^2}{m_o c^2} \qquad (3b)$$

It is seen that the appearance of the new hidden dimension occurs in a form analogous to the splitting of a canceling charge pair of particles from the vacuum, by splitting of a quantized light-like, or vacuum, space-time interval of length zero. In the GEM theory the hidden dimension size, where the hidden dimension can mix with the non-hidden dimensions, is the quantized particle size. The hidden dimension quantities are thus able to mix with the normal spacetime quantities because they are similar at small scales. This will lead to, as we experience them, two particle types. One is associated with the time-like portion of the constrained interval, leading to a one-dimensional character, an electron, and another of equal size with a space-like character having three constrained sub-dimensions, a proton. Despite the three sub-dimensional character of the space-like portion, the symmetry with the time-like portion of the split interval causes the hidden dimension to behave as a one dimensional quantity. It will be seen that the quantity we call particle charge is a geometric distance within the constrained spacetime interval. Thus, the formation of the hidden dimension manifests itself as the image of space-time, but split into a space-like portion, the proton, and a time-like portion, the electron. The appearance of charge in the vacuum, because of the presence of the vacuum quantity G, gives us a distance as seen in Eq. 4. Thus G functions in the vacuum as the "interpreter" of charge into either mass or distance. Thus, ironically, charge and mass, the source terms for EM and gravity, are unified already in the vacuum quantity G, which has units of charge to mass ratio squared in the esu system used here.

$$q\sqrt{G/c^4} = r \qquad (4)$$

Therefore the quantized vacuum scale length, the Planck length, gives birth to a quantized larger scale hidden dimension. Because the quantized hidden dimension is an image of macroscopic space-time in a light-like interval, and its structure is part of a split "lightlike" spacetime where charge, q, is analogous to macroscopic dimensions we have charge conservation and interval conservation. We have we have the following constraints on the charges of the particles,

$$q_o = -q_t = q_x + q_y + q_z \qquad (5a)$$

$$q_o^2 = q_t^2 = q_x^2 + q_y^2 + q_z^2 \qquad (5b)$$

where the subscripts denote the corresponding time or space dimensions in the unconstrained Cosmos. Thus the space-like portion of the split interval, the proton, has three sub-dimensions that we interpret as quarks or sub-charges, while the electron acts like a single entity.

Beyond Einstein: GEM Unification

II. The Covariant Formulation of the GEM theory

Sakharov began his exploration of the relationship between EM and gravity with Hilbert action principle, where the vacuum equations of Einstein's theory of general relativity from the extremization of the action integral:

$$W = (16\pi G)^{-1} \int (R - 2\Lambda)\sqrt{-g}\, dx^4 \quad (6)$$

where R is the Ricci curvature scalar of General relativity, Λ is the cosmological constant and G is the Newton Gravitation constant. In the concept of Sahkarov, the Λ term represented a zeroth order background EM field whereas R term represented first order perturbation on this field caused by space-time curvature. This integral was used by Sahkarov to relate the force of gravity to a radiation pressure, or Poynting flux, due to the ZPF. The ZPF arising naturally from the Heisenberg uncertainty principle. In his article, Sahkarov derived the value of G, as the "metric elasticity of space" by assuming that the Planck Length would form the short wavelength cutoff for the ZPF spectrum, and that space-time curvature formed a perturbation on this spectrum. This led to the result of Sahkarov that gravitation constant can be found as an integral over the ZPF spectrum:

$$G^{-1} \cong \frac{\hbar}{2c^5} \int_0^{\omega^*} \omega\, d\omega = \frac{\hbar \omega_P^2}{c^5} \quad (7a)$$

$$G = \frac{c^3 r_P^2}{\hbar} = \frac{c^4}{r_P^2 T_o} \quad (7b)$$

where the energy density $T_o = \hbar c / r_P^4$ is the Planck scale energy density.

Beyond Einstein: GEM Unification

As was shown in the first GEM article (Brandenburg 1991) an ExB or Poynting drift field, with constant B and E growing stronger in the direction of the drift, can produce a gravity like acceleration of charged particles of all charges and masses, as shown in Figure 1. The Sakharov model for the gravity force is basically that of a radiation pressure Poynting field produced by non-uniformities in the ZPF and is successful in the sense that is self consistent. However, what is needed is a formula for the Planck length that is independent of G. We can begin a covariant formulation of the GEM theory from the expressions in the first GEM article, where the ZPF stress energy was caused to vanish (Brandenburg 1991):

$$g_{\alpha\beta} = \frac{4 F_\alpha^\gamma F_{\gamma\beta}}{F_{\mu\gamma} F^{\mu\gamma}} \quad (8a)$$

$$T_{\alpha\beta} = F_\alpha^\gamma F_{\gamma\beta} - g_{\alpha\beta} \frac{F_{\mu\gamma} F^{\mu\gamma}}{4} = 0 \quad (8b)$$

The key principle is then that the metric tensor and the stress tensor, both of which are symmetric, are proportional in very strong fields. When this happens, the fields become geometry and vanish. The universe thus indulges in *self-censorship*. This is necessary for us to enjoy a low temperature Cosmos. For the case of weak fields perturbing this self censorship we have the gravity potential in the Newtonian limit as a perturbation on the metric formed from ultra strong fields.

We have for a perturbed tensor

Beyond Einstein: GEM Unification

$$g_{\alpha\beta} = \frac{|F_\alpha^{\ \gamma} F_{\gamma\beta}|}{F_{\mu\gamma} F^{\mu\gamma}} = (B^2 - E^2)^{-1} \begin{pmatrix} -E^2 & S_x & S_y & S_z \\ S_x & E_x^2 - B_y^2 - B_z^2 & E_x E_y + B_x B_y & E_x E_z + B_x B_z \\ S_y & E_x E_y + B_x B_y & E_y^2 - B_x^2 - B_z^2 & E_y E_z + B_y B_z \\ S_z & E_x E_z + B_x B_z & E_y E_z + B_y B_z & E_z^2 - B_y^2 - B_x^2 \end{pmatrix} \quad (9a)$$

This yields, at the Planck Scale, metrics of the form

$$g_{\alpha\beta} = \begin{pmatrix} -1 & 0 & 0 & 0 \\ 0 & 1 & 0 & 0 \\ 0 & 0 & 0 & 0 \\ 0 & 0 & 0 & -2 \end{pmatrix} \quad (9b)$$

We apply the condition that the particles choose the shortest path through spacetime, and the ZPF is isotropic on large scales, so, when averaged over a region of spacetime, is isotropic, uniform and is dominated by magnetic flux:

$$E^2 = \frac{B^2}{2} \quad (10a)$$

so that, for geodesics, an effective Minkowski space arises when averaged at large scale.

$$g_{\alpha\beta} = \begin{pmatrix} -1 & 0 & 0 & 0 \\ 0 & 1 & 0 & 0 \\ 0 & 0 & 1 & 0 \\ 0 & 0 & 0 & 1 \end{pmatrix} \quad (10b)$$

313

Beyond Einstein: GEM Unification

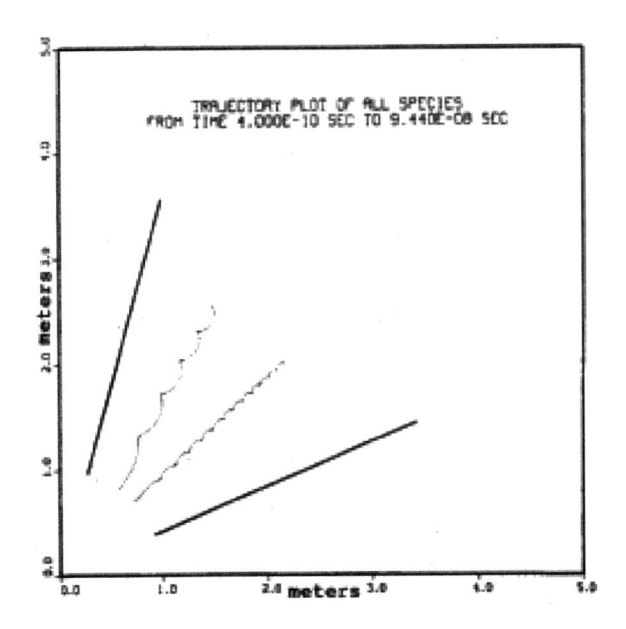

Figure 1. Charged particles, an electron and heavy positron (10x m_e) move in a simulated GEM gravity field.

Thus we can incorporate the Sakharov concept of spacetime as due to EM field into a theory of the metric tensor as apportion of the EM stress tensor and when this is done the full stress tensor vanishes. Thus we can have a spacetime that is fabric of EM energy that conceals itself. If we extremize the action, we obtain the vacuum equations of General Relativity

$$R_{\mu\nu} - \frac{1}{2} g_{\mu\nu} R = g_{\mu\nu} \Lambda + 8\pi \frac{G}{c^4} T^o_{\mu\nu} =$$

$$g_{\mu\nu} \Lambda + 8\pi \frac{G}{c^4} (F^\alpha_\mu F_{\alpha\mu} - F^\alpha_\mu F_{\alpha\mu}) \qquad (11)$$

Beyond Einstein: GEM Unification

where the EM stress tensor T°$_{\mu\nu}$ consists of two canceling portions. However, we see that the two canceling portions can be written, with the condition

$$F_{\alpha\beta}F^{\alpha\beta} = \frac{hc}{8\pi r_P^4} \qquad (12)$$

$$g_{\mu\nu}\Lambda + 8\pi \frac{G}{c^4}(F_\mu^\alpha F_{\alpha\mu} - F_\mu^\alpha F_{\alpha\mu})$$
$$= g_{\mu\nu}\Lambda + g_{\mu\nu}\frac{1}{r_P^2} - g_{\mu\nu}\frac{1}{r_P^2} = g_{\mu\nu}\Lambda \qquad (13)$$

Therefore the two canceling portions of the spacetime stress tensor assume the form of cosmological constant terms before they cancel. This is a mechanism for the cancelation of the ZPF via a cosmological constant that was first proposed by Zeldovich (1967). Thus our expression for the metric as being a normalized portion of the EM stress tensor appears valid for the case of spacetime without particles or their fields. This gives a reasonable description of the vacuum, but what of the metric in the presence of fields and their source particles? To transition from the vacuum state we must allow a new degree of freedom to appear, which is the fifth dimension of Kaluza and Klein (1929)

The Kaluza-Klein Fifth Dimension

The Standard Model describes a cosmos where particles and both EM fields and gravity fields exist, and the only theory which derives both sets of fields from the same principle is the Kaluza-Klein theory involving a "hidden" 5th dimension. The Kaluza and Klein theory extended the Hilbert action principle, from which GR can be derived, by adding an additional dimension that is "compactified" to small fixed length "ε" so that we have for an action principle for a universe full of hydrogen and gravity and EM fields:

$$W' = (16\pi G)^{-1} \int K\sqrt{-g}\, dx^5 \qquad (14)$$

$$\gamma_{ab} = \begin{bmatrix} g_{\tau\upsilon} + \xi\kappa^2 A_\tau A_\upsilon & \xi\kappa A_\tau \\ \xi\kappa A_\upsilon & \xi \end{bmatrix} \qquad (15)$$

where the energy density K can be written, in the limit of the new dimension being compactified

$$K = R - 2\Lambda + \kappa^2 \left[\xi \left[\frac{E^2 - B^2}{8\pi} \right] + \gamma_{ab} \Theta^{ab} \right] \qquad (16)$$

where E and B are the electric, and magnetic field strengths, and $\kappa^2 = 16\pi G/c^4$. Where we have defined the metric tensor over dimensions 1-5 as a or b and dimensions 1-4 of conventional space-time as the Greek letters τ or υ. We note that element $\gamma_{55} = \xi$ is the element controlling the length of the line element in the fifth dimension with itself, so that for ξ=0 the fifth dimension does not exist and for ξ=1, it is fully deployed. Following the original formalism (Klein 1926), we have a generalized momentum-stress tensor due to particle motion:

Beyond Einstein: GEM Unification

$$\Theta^{ab} = \frac{\left[\sum_n \gamma_{ab} \frac{dx_p^a}{d\ell} \frac{dx_p^a}{d\ell} + \sum_n \gamma_{ab} \frac{dx_e^a}{d\ell} \frac{dx_e^a}{d\ell}\right]}{V} \qquad (17)$$

where we have position vectors x_p^a and x_e^b, for protons and electrons, respectively and density is defined as a local sum over a small volume, V.

In a geometric interpretation, the signature of the coupled appearance of particles and EM fields in the GEM theory is the relation between curvature radius of the Planck Length $r_P = (G\hbar/c^3)^{1/2}$ and what is termed the "mesoscale" radius, typical of classical particle radii, r_o.

$$r_o = \frac{e^2}{m_o c^2} \qquad (18)$$

This is based on the postulate that baryon and lepton number disappear at the Planck scale coincidentally with the separate identity of Gravity and EM fields. The appearance of the Kaluza-Klein fifth dimension is thus the appearance of particles- electrons and protons with their classical radii. This can be seen intuitively from a Gedanken experiment described in Brandenburg (1995), where a single atom of hydrogen is collapsed to the Planck scale, forms a black hole and upon Hawking evaporation, reappears as a shower of particle and antiparticles, thus destroying baryon and lepton number.

In the GEM model both protons and electrons begin in a U(1) symmetric field, the simplest possible field symmetry consistent with QED, for a mesoscale "union" particle with

Beyond Einstein: GEM Unification

mass $m_o = (m_p m_e)^{1/2}$. This union field exists at the Planck length

$$m = m_v \cos(\phi) + i m_n \sin(\phi) \qquad (19)$$

The angle ϕ, we will consider in this model corresponds to charge state and is thus quantized as a canceling pair $\pm\phi_o$, even in the Planck Scale. The appearance of the compact fifth dimension allows this angle to become an imaginary rotation angle to give two real particle masses corresponding to an "up" quantum state and "down" quantum state from the U(1) symmetry

$$m = m_o \exp(\pm\phi_o(\xi)) \qquad (20)$$

So that separate particle masses are generated at $\xi=1$ from Eq. 20 by

$$\sigma = \exp(\phi_o(\xi)) \qquad (21)$$

So that even though mass symmetry is broken in terms of the new 5 space we experience, it is preserved in terms of a geometry involving the imaginary angles in the original U(1) symmetry. That is, the new particle dimension looks symmetric in the space of imaginary angle.

To obtain a smooth transition to the union field as curvature collapses to the Planck scale the angle ϕ_o must be dependent on curvature near the Planck length but very insensitive to it at larger curvatures, where the new fifth dimension is fully developed. Based on the lack of observation of proton decay, lepton and baryon numbers are obviously strongly conserved. The simplest way to obtain this

Beyond Einstein: GEM Unification

mixture of scale sensitivity with curvature r_c is for the rotation angle to have the dependence

$$\phi_o(\eta) = \ln(\ln(r_c(\xi)/r_p)) \qquad (22)$$

with

$$r_c(\xi) = \xi r_o \qquad (23)$$

so $\xi=0$ is 4-space and $\xi=1$ is 5-space.
Therefore, in the GEM model, the separate appearance of protons and electrons is, like the separate appearance of EM and gravity forces, linked to the appearance and full development of the Kaluza-Klein fifth dimension. The physical description of this dimension is thus as a particle classical radius r_o. The expansion of the effective curvature to r_o, the mesoscale radius of classical particle radii then yields

$$\frac{dx_i^a}{d\ell} = \frac{dx_i^a}{d\tau}\frac{d\tau}{d\ell} = \frac{dx_i^a}{d\tau}\sqrt{m_i} \qquad (24a)$$

$$\frac{dx_i^0}{d\tau} = \frac{e}{\kappa c m_i} \qquad (24b)$$

where i represents either p or e , we obtain then for Eq. 17, with the substitution $m_o c^2 = e^2/r_o$

$$\Theta^{ab} = \frac{\dfrac{e^2}{r_o c^2}\left[e^{\phi_o}\sum_n \gamma_{ab}\dfrac{dx_p^a}{d\tau}\dfrac{dx_p^a}{d\tau} + e^{-\phi_o}\sum_n \gamma_{ab}\dfrac{dx_e^a}{d\tau}\dfrac{dx_e^a}{d\tau}\right]}{V} \qquad (25)$$

Beyond Einstein: GEM Unification

and we define a local density n_i and a mean position $\bar{x}_i$ so that we have for our new set of equations in a hydrodynamic form

$$\frac{\partial_\mu F^{\mu\nu}}{\kappa} = \frac{e}{\kappa c} n_p \frac{d\bar{x}_p^\nu}{d\tau} - \frac{e}{\kappa c} n_e \frac{d\bar{x}_e^\nu}{d\tau} \qquad (26)$$

$$R^{\mu\nu} - \frac{1}{2} g^{\mu\nu} R = \kappa^2 T^{\mu\nu} + \kappa^2 \frac{e^2}{r_o c^2} \left[e^{\phi_o} n_p \frac{d\bar{x}_p^\nu}{d\tau} \frac{d\bar{x}_p^\mu}{d\tau} + e^{-\phi_o} n_e \frac{d\bar{x}_e^\nu}{d\tau} \frac{d\bar{x}_e^\mu}{d\tau} \right] \qquad (27)$$

Thus we recover both the non-vacuum Maxwell's equations and the equations of General Relativity for a hydrogen plasma with the hidden fifth dimension size seen explicitly.

This constitutes a very basic description of the Cosmos as a whole, whose major known particle constituents are protons and electrons. If one conceives that the action principle assumes a massless EM field then the appearance of the fifth dimension allows trapping or scattering of massless quanta to create rest mass, charge, and hence particles.

We have then for perturbing fields for the portion of the metric tensor for fields around a charged particle.

$$g_{00} = 1 - 2\phi = \frac{E^2}{B^2 - E^2}, \quad E^2 = E_0^2 + E_1^2 \qquad (28)$$

$$1 - 2\phi = 1 - \frac{E_1^2}{B^2} \qquad (29)$$

$$\frac{\partial V_D}{\partial t} = V_D \frac{\partial V_D}{\partial x} = \frac{E}{B} c \frac{1}{B} \frac{\partial E}{\partial x} c \qquad (30)$$

Beyond Einstein: GEM Unification

Quantum Field Perturbation Model

We will leave for a later article, the development of a full quantum field description of what we have done thus far. However, we must acknowledge at this point the quantum nature of this theory of unification which springs from the Planck scale quantities where space-time itself is quantized in the Planck length. The particles and equations are the result of an extremum of action, and thus variations around the extreme must exist and give rise to higher order quantum phenomenon in the weak field limit. Thus a robust perturbation scenario must exist.

In a system of an electron and proton interacting, such a hydrogen atom, gravity and EM perturbations exist and affect the observed frequencies. In the GEM theory we can have proton-electron exchange perturbations, that is, the electron and proton can temporarily exchange places. This remarkable effect will hide itself in the commonplace, just as electrodynamics hides itself in the vacuum, and we will assume it is already been accounted for in the "reduced mass formalism" ,of classical mechanics where the masses of the co-interacting particles are made to disappear explicitly and the correction term $(1+m_e/m_p)^{1/2}$ appears on the frequency to account for it. Here we will make the assumption, that at least to leading order, this lepton-baryon exchange perturbation can be accounted for by the reduced mass correction.

$$correction\ \ terms \cong (1+m_e/m_p)^{1/2} = 1+1/\sigma^2... \qquad (31)$$

Mass and Charge From the Vacuum

Based on the postulate that baryon and lepton number disappear at the Planck scale, and do this coincidentally with

Beyond Einstein: GEM Unification

the separate identity of Gravity and EM fields, the appearance of the Kaluza-Klein fifth dimension is thus the appearance of particles-electrons and protons with their classical radii. The expansion of the effective curvature to r_o, the mesoscale radius of classical particle radii then yields (Brandenburg 2010)

$$\left(\frac{m_p}{m_e}\right)^{1/2} = \ln\left(r_o/r_p\right) = \frac{\sigma}{1+C_2/\sigma^3} \quad (32)$$

Where $C_2 = 0.86$. However, the MIT bag model (Kodos 1978) gives a relationship for σ from a calculation of the energy of three confined quarks.

$$\sigma \cong K\alpha^{-1/2} \quad (33)$$

In GEM theory, charge is length in a hidden dimension, so it is not surprising that distance in the hidden dimension is not only restricted but quantized.

We can understand the emergence of charge and mass from the vacuum as an extension of the exponential relationship between charge and mass for the proton-electron pair.

$$\ln\left(\frac{m}{m_o}\right) = \frac{q}{e}\ln\sigma \quad (34)$$

where the charge of particle q is $\pm e$. We can also write, because $m_o/M_p =$

$$\frac{M_P}{m_o} = \frac{r_o}{r_P}\frac{\hbar c}{e^2} \quad (35)$$

$$\ln\left(\frac{M_P}{m_o}\right) = \sigma + \ln\alpha \quad (36)$$

By extension, we have the appearance of the electron charge from the vacuum itself

Beyond Einstein: GEM Unification

$$\alpha^{-1/2} = \left(\frac{\hbar c}{e^2}\right)^{1/2} = \frac{q_v}{e} \tag{37}$$

It would then be logical that for the logarithmic relationship between charge and mass of Eq. 34 to be extended, the mass of the proton, which carries all the mass of the electron-proton system, should be logarithmically related to the Planck mass by the ratio of the "Planck charge", to the particle scale charge:

$$\ln(\frac{M_P}{m_p}) \cong \alpha^{-1/2}(1+1/\sigma^2)\ln\sigma \tag{38}$$

This relationship indeed exists numerically. Thus, the appearance of the proton from the vacuum mass M_P is related logarithmically to the corresponding appearance of the electron charge from the vacuum charge $\hbar c$. We can extend this to the relationship between the mesoscale mass and the Planck mass, and be extension the ratio of the mesoscale radius to the Planck length:

$$\ln(\frac{M_P}{m_o}) \cong \ln\sigma(\alpha^{-1/2}(1+1/\sigma^2)+1) \tag{39}$$

We can then by Eq.s 32 and 36 obtain a transcendental equation for σ itself:

$$\ln(\frac{r_o}{r_P}) = \frac{\sigma}{\left(1+\frac{C_2}{\sigma^3}\right)} \cong \ln\sigma(\alpha^{-1/2}(1+1/\sigma^2)+1) - \ln(\alpha^{-1}) \tag{40}$$

Given the operation of ratio of masses to m_o is involved we must admit the presence of quantum exchange in this operation of second order, thus we approximate to leading order in mass ratio, similar to Eq. 33:

$$\sigma \cong (\ln\sigma(\alpha^{-1/2}(1+1/\sigma^2)+1) - \ln\alpha^{-1})(1+0.86/\sigma^3) \tag{41}$$

Beyond Einstein: GEM Unification

solving this transcendental equation results in the value

$$\sigma = 42.8503..., \sigma^2 = 1836.15 \quad (42)$$

The gravitation constant can be derived to zeroth order from Eq. 10, and following this form the gravity interaction in the macroscopic world can be written, for two masses M_1 and M_2, separated by distance d, as

$$F = \frac{GM_1M_2}{d^2} \cong \frac{e^2}{d^2}\frac{M_1}{m_o}\frac{M_2}{m_o}\alpha\exp(-2\sigma)(1+0.86/\sigma^2...) \quad (43)$$

so that the gravity interaction between two masses can considered a "residual" EM interaction where the two masses are each normalized by m_o and then assigned an elementary charge e according to this norm, the resulting EM interaction is then assigned a probability $\exp(-2\sigma)$ times the EM probability α of the interaction. The gravity interaction is thus three GEM interactions, two norms and a scaled EM interaction. We can assume that lepton-baryon exchange effects each of these basic interactions and thus assign a factor of $(1+\sigma^{-2})^{-1/2}$ to each to approximately account for the weakening each interaction due to electron-proton quantum swapping. We thus obtain for the gravity constant, using the measured value of the proton electron mass ratio, to first order:

$$G = \left(e^2/m_p m_e\right)\alpha\exp(-2\sigma)(1+0.86/\sigma^2....)$$
$$= 6.67424 \times 10^{-8} dyne-cm^2 gm^{-2} \quad (44)$$

324

Beyond Einstein: GEM Unification

this expression is accurate to within .8 parts per ten thousand and thus is within experimental error.
$G = 6.67428 \times 10^{-11}$ m³ kg⁻¹ s⁻² $= 6.67428 \times 10^{-11}$ N (m/kg)².

III. Quarks and the Subdimension of the 5ᵗʰ Dimension and the GEM Strong Theory

If there is any portion of the Standard Model the GEM theory appears in conflict with, it is the GEM theory's treatment of the proton as a fundamental particle, instead of a composite object formed of quarks. This however, is not a serious problem. For the Kaluza-Klein fifth dimension can be considered to be constrained image of spacetime and thus having four sub-dimensions. The electron corresponds to a "time-like" or scalar entity while the proton corresponds to a space-like component, having three sub-dimensions. We can minimize the volume of this dimension, given the two constraints of charge conservation and the conservation of mesoscale radius, defined in Eq. 9, which is a constraint on the sum of the squares of quark charges.

$$q_1 q_2 q_3 + \lambda_1(q_1^2 + q_2^2 + q_3^2) + \lambda_2(q_1 + q_2 + q_3) \tag{45}$$

$$q_1^2 + q_2^2 + q_3^2 = 1 \tag{45}$$

$$q_1 + q_2 + q_3 = 1 \tag{46}$$

$$q_1 q_2 + 2\lambda_1 q_3 + \lambda_2 = 0 \tag{47}$$

$$q_1 q_3 + 2\lambda_1 q_2 + \lambda_2 = 0 \tag{48}$$

$$q_3 q_2 + 2\lambda_1 q_1 + \lambda_2 = 0 \tag{49}$$

which has the solutions

$$\lambda_1 = \frac{1}{3} \quad \lambda_2 = -\frac{2}{9} \qquad (50a)$$

$$q_1 = -\frac{1}{3}, q_2 = q_3 = \frac{2}{3} \qquad (50b)$$

This corresponds to the standard quark model, and the second solution is that of an electron with q/e = -1 and . Thus in solving the problem of the structure of a 5th dimension, one finds that its volume, upon being minimized, with constraints, yields the charges of the quark system. Thus, we can now call the GEM theory a GEM-strong force theory.

IV. Discussion: GEM and the Standard Model

The basis for applying a quantum description of matter and fields can be found in the quantum character of lengths masses and charge-action in the vacuum. The two long range forces EM, and gravity, can be related, at least in terms of their relative coupling strengths and length scale by a model of the emergence of quantum particles and their fields from the vacuum. They all separated from each other with the birth of a new dimension, which carried with it and inherent quantum scale length that is the signature of matter in the Cosmos. This scale length and associated quantities of charges and masses can be related to the Planck Scale also through the Standard Model by the observed variation of α:

Beyond Einstein: GEM Unification

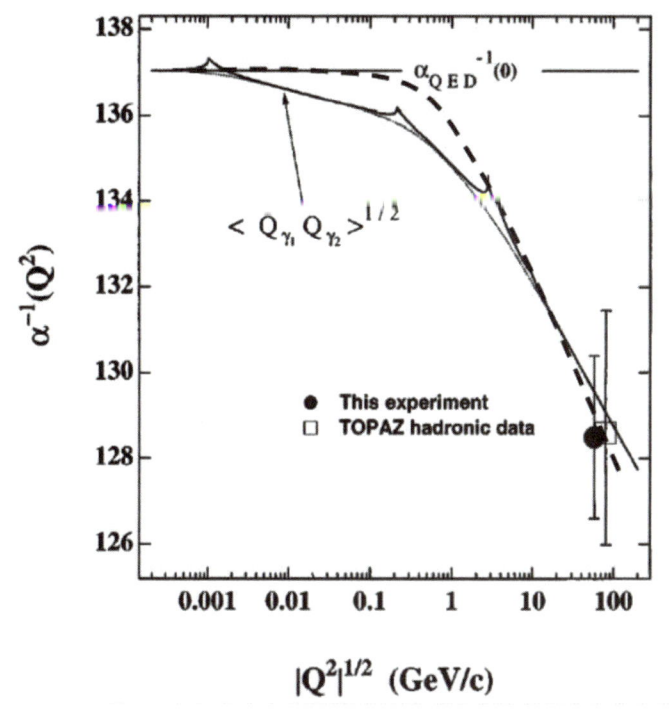

Figure 2. The variation of α with energy, with a model calculation (dashed line) fit to zero energy, and 10 GeV.

The variation of alpha with energy (see Figure 1.) can be approximated in the region of 0 to 10 Gev from its curve of variation as the function

$$\alpha^{-1} \cong 3.20\ln\left[\frac{W^* + w_{probe}}{W_P}\right] \quad (51)$$

where $W_P = \hbar c/r_P$ is the Planck energy, and $W^* = 3.0\text{GeV} \cong \hbar c/r_o$ is the approximate mass energy $M_{\eta c}$ of the η_{cc} lowest lying state of Charmonium, and w_p is the probe energy. This also nearly the energy of the $\Sigma(3000)$ resonance. Thus the value of α^{-1} asymptotes to 137 at $w_p \ll W^*$. Since the

Beyond Einstein: GEM Unification

expression in Eq. 17 involves the Planck energy, it can be inverted to find an approximate expression for G :

$$G \cong \frac{\hbar c}{M_{\eta\mu}^2} \exp\left[\frac{-1}{1.599\alpha}\right] = 7.10 \times 10^{-8} dyne - cm^2 g^{-2} \qquad (52)$$

This expression is very similar to that derived by M.J. Clarke (2003), based on a quantum model of gravity forces as being due to boson exchange.

$$\frac{e^2}{m_p m_e} \alpha = \frac{\hbar c}{M_{\eta c}^2} \qquad (53a)$$

$$\sigma = (3.198\alpha)^{-1} = \left(\frac{m_p}{m_e}\right)^{1/2} = 42.8503...... \qquad (53b)$$

The GEM theory gives us some useful basic concepts but it obvious that is only an approximate description of a Cosmos which is much richer and more detailed. A glaring example of this that the proton and electron yield the same size, as predicted by GEM, but the size : 1.4fm, obtained when the electron is treated as a charged sphere and its mass is derived from the electrostatic field, and the proton size is analyzed is derived from nuclear volumes. However, this size is not the size suggested by GEM, r_o, which is much smaller. The GEM also does not describe even basic particles such as neutrons, until they have decayed into electrons and protons and then can say nothing about the resulting neutrino. Thus there is much room for improvement.

Beyond Einstein: GEM Unification

The basic results flowing from the GEM theory are that gravity can be understood as collection of ExB drift cells formed in the ZPF and that protons and electrons are the space-like and time-like parts respectively of a zero length interval. The square root of mass ratio, so important to this theory, is also seen as being simply related to the fine structure constant in approximately the same fashion as the MIT Bag Model. The fundamental relations of the logarithmic relation between the ratio of the classical particle size and Planck length and the ratio of a particle mass to some standard mass can be understood as a validation of renormalization, which itself indicates that the Planck length forms a cutoff at short wavelengths.

If the GEM theory validates anything about the approaches of mainstream physics it is in tracing the origin of quantities associated with matter back to Planck scale quantities r_p and M_p, thus turning them from hypothetical limits of integration into hard values associated with everyday quantities. In particular the ubiquitous procedure of re-normalization where singularities in field theories become logarithmic due to quantum effects is validated and grounded in its ultraviolet limit in the Planck scale. Thus physics has arrived at the former frontier of the Planck scale.

The relationship between gravity and EM is may be tied to the properties of the lowest mass Charmonium meson , which plays a prominent role in parity violation experiments and may lie close to the mass of the elusive Dark Matter particle. It now appears possible that a fully covariant theory of the GEM theory is possible, and while this effort has just begun, its results are encouraging. Such a covariant GEM theory, may have profound implications for superluminal drive and communications.

NOMENCLATURE

A =area (m^2)
A=acceleration (m/s^2)

Beyond Einstein: GEM Unification

A_v = EM four potential, (Volts)
e = electronic charge(Coulombs) ,
E = electric field(Volts/m),
B = Magnetic field(Webbers/m^2),
c = speed of light(m/s),
G = Newton gravitation constant(kg^{-2}-m^3-s^2) ,
$\sqrt{-g}$ = square root of the determinate of the metric tensor (pure number),
$g_{v,u}$ = metric tensor (pure number),
g.= Newtonian gravity acceleration (m-s^{-2}) ,
h = Planck's constant (kg-m^2-s^{-1}),
$\hbar$ = Planck's constant divided by 2π (kg-m^2-s^{-1}),
L= Lagrangian (kg-m^2-s^{-1})
L'= Lagrangian (kg-m^2-s^{-1})
K= energy density (Joules/m^3)
u= energy density (Joules/m^3)
m_p = proton mass (kg),
m_e =electron mass (kg),
S = Poynting flux (Watts-m^{-2}).
R = Curvature scalar (m^{-2}),
R_m=Proton to electron mass ratio (pure number)
r_c=classical particle radius (m)
r_o=equilibrium radius
V_D= drift velocity (m/s)
v=velocity (m/s)
α=fine structure constant (pure number),
α_S = strong force fine structure constant (pure number),
ϕ = wave field potential (m$^{-3/2}$)
γ_{ab} = five dimension metric tensor,
Λ = Cosmological constant, (m^{-2})
ρ_m =mass density (kg/m^3)
$\xi = \gamma_{55}$
ε_o=permitivity of space (m^{-1}-coulomb^2/Joule)
μ_o=permeability of space (Joule-s^2/Coulomb^2)

Beyond Einstein: GEM Unification

ACKNOWLEDGMENTS

The author wishes to thank Ron Phillips, Jaydeep Mukherjee, and Robert Crabbs, of the Florida Space Institute, and Eric Rice of Orbital technologies Corporation for their encouragement of this research. The author extends special thanks to Morgan Boardman and Paul Murad of Morningstar LLC for their financial support of this research.

REFERENCES

Brandenburg, J. E., "Unification of Gravity and Electromagnetism in the Plasma Universe", IEEE *Transaction in Plasma Science*, 20, 6, 1992, pp. 944

Brandenburg, J. E., "A Model Cosmology Based on Gravity-Electromagnetism Unification", Astrophysics and Space Science, 227, 1995, pp. 133

Brandenburg, J.E. (2007) "The Value of the Gravitation Constant and its Relation to Cosmic Electrodynamics," IEEE Transactions On Plasma Science, Plasma Cosmology Issue Vol. 35, No. 4., p845.

Brandenburg J.E. and Kline J., Joint Propulsion Conf. AAIA-98-3137, (1998).

Brandenburg, J. (2010) "A Combined Kaluza-Klein Sakhaov Model of Baryo-Genesis" Ohio Chapter of the APS meeting Flint Mich.

Chen F. F., Introduction to Plasma Physics, Plenum Press NY,(1976), pp. 21.

Clark M.J. " A Theroretical Formula for the Gravitation Constant Based on a Boson Theory" from The Gravitation Constant: Porceedings of a NATO Summer Study, Granazzli H.and Gilley A. Editors (2003)

Haisch, B., Rueda, A., and Puthoff, H.E., "Advances in the Proposed Electromagentic Zero-Point Field Theory of Inertia", AAIA 98-3143, (1998).

Klein, Oskar "Quantum Theory and Five-Dimensional Relativity Zeitschrift fur Physik, 37, 895, (1926).

Jackson J.D., "Classical Electro-Dynamics" Wiley and Sons, New York, 1975, pp. 205.

Kozyrev, N., A., Joint Publications Research Service # 45238, (1968).

Levine I. et al "Measurenment of the Electromagnetic Coupling at Large Momentum Transfer" Phys. Rev. Lett. 78, 3, (1997)

Puthoff, H.E., "Gravity as a zero-point fluctuation force", *Phys. Rev.* **A 39**, 2333, (1989).

Misner C.W., Thorne, Kip S., and Wheeler J.A., "Gravitation", W. H. Freeman and Company, San Francisco, 1973, pp. 465

Sakharov A.D. "Vacuum quantum fluctuations in curved space and the theory of gravitation" Sov. Phys. Doklady 12,1040-1041, (1968).

Sen , A, "Remarks on Tachyon Driven Cosmology" Physicia Scripta Vol. T117, 70-75, 2005.

Zel'dovich, Ya. B., Cosmological constant and elementary particles, Sov. Phys.--JETP Lett. 6,316-317, (1967).

Beyond Einstein: GEM Unification

Index

anti-matter, 37-40, 51, 169-171, 227, 233, 258
Aristotle, 25, 27, 66, 67, 75, 92, 138, 147
Aristarchus, 67
Bohr, Niels, 109-115, 119-120, 128, 275, 306
Chandrasekar, Sudyam, 141, 154, 155
Cook, Nicholas, 252
Davis, Eric, 233
Democritus, 3, 22
Dirac, Paul, 36-42, 118, 167, 169, 263-264, 273, 290, 291, 298
Dirac Sea, 38
Dirac large numbers hypothesis, 38, 398
Eddington, Arthur, 140, 141
Eratosthenes, 24, 41, 130
Faraday Michael, 81, 85-86, 90, 101
Faraday Tensor, 183, 188, 190, 191
Farrell, Joseph, 252
Feynman, Richard, 115-118, 122, 204, 275-276
Freidman, Stanton, 252, 280
Franklin, Benjamin, 84,85
Gell-Mann, Murray, 202,285
Gilbert, William, 69,83
Greene, Brian, 231
Guth, Allen, 257
Grand Unified Theory (GUT), 204
Haisch, B., 276
Hawking, Steven, 171-172, 217, 273
Hawking evaporation, 217, 262, 266, 275, 317
Hilbert David, 113-114, 131, 170

Beyond Einstein: GEM Unification

Hilbert Action Principle, 132, 137, 144, 196, 217, 244, 311, 315
Heisenberg, Werner, 112-115, 160-169, 228, 232
Heisenberg Uncertainty, 38, 160, 213, 217, 311
Hellings, Ronald, 42, 264
Hubble Time, 26, 263, 264, 266, 283, 288
Hubble radius, 27, 37, 41-42, 263-264, 284
Huygens Christian, 72-74, 105, 107
Imaginary numbers, 60, 254, 269-270, 295
Inverse Square Law, 70, 76-77, 79, 83, 99, 116, 164-165, 181, 190, 201, 267
Kepler, Johannes, 65, 68-70, 75, 76
Kaku, Michio, 277
Kaluza, Theodore, 144, 196-199
Kaluza-Klein Theory, 199, 201-202, 204-205, 218, 230, 232, 258, 273, 277, 279,-280, 284, 297
Klein, Oskar, 144, 199, 200
Edward P. Lee, 218
Lodestones, 63, 82, 91, 93, 101
Loop Gravity, 224, 276
Maxwell, James Clerk, 25, 81, 86-93, 98-101, 103, 105, 113, 117, 122
Maxwell stress tensor, 197-198, 276
Metric, 182, 186, 195, 222, 276, 305
Mesoscale, 208, 218, 227, 229, 296
Newton, Isaac, 6, 65, 71-72, 75, 76-80, 98, 101, 105-108, 111, 138-139, 155, 172, 219, 230, 244, 275
Newton Gravitation Constant, 172, 219, 237, 230-233, 244, 264, 274, 279, 287
Nordstrom, Gunnar, 119, 195, 273
Oppenheimer, Robert, 119, 155-157, 159, 160, 161

Beyond Einstein: GEM Unification

Pauli, Wolfgang, 37, 40, 119, 144, 185, 201, 202
Planck, Max, 35, 103-108, 119, 125-128, 140, 164, 166
Planck scale , 177, 214, 224-225, 227-229, 234 235
Planck charge, 229
Planck length, 172, 176, 177, 222, 226, 231, 232, 234, 237, 238-239, 268, 273, 278, 286, 289
Planck-mass energy, 211,177
Poynting, John Henry, 91,95, 99,
Poynting field, 95,99-100,122,155-156, 170
Poynting force,101, 171, 176, 214, 220, 243, 258, 285-286, 290
Proton decay, 211
Puthoff, Hal, 286
Real numbers, 53
Reimann, Bernhard, 133-135, 138-139
Rueda, Alfonso, 276
Sakharov, Andrie, 159-163,165-171, 208, 227, 231, 244, 258, 273, 276-277
Szilard, Leo, 157-158,180
Smolin, Lee, 224-225, 276
T'Hooft, Gerald, 172-173
Teller, Edward, 157-158, 163
Tesla , Nikola, 4-6, 11-12, 122,124, 175-178, 179-184, 189-193, 247-251,273, 277, 280
Tesla Vortex, 247-251, 280
Thales, 25, 82,
TOE (theory of everything) , 204
Vacuum Decay, 206, 259, 261-267,
Williamson, Lance, 265
Witten, Edward , 206, 258, 259, 283
Zeno, 22, 24, 175, 192, 273
Zeno's Paradox, 192,

Get these fascinating books from your nearest bookstore or directly from: Adventures Unlimited Press
www.adventuresunlimitedpress.com

DEATH ON MARS
The Discovery of a Planetary Nuclear Massacre
By John E. Brandenburg, Ph.D.

New proof of a nuclear catastrophe on Mars! In an epic story of discovery, strong evidence is presented for a dead civilization on Mars and the shocking reason for its demise: an ancient planetary-scale nuclear massacre leaving isotopic traces of vast explosions that endure to our present age. The story told by a wide range of Mars data is now clear. Mars was once Earth-like in climate, with an ocean and rivers, and for a long period became home to both plant and animal life, including a humanoid civilization. Then, for unfathomable reasons, a massive thermo-nuclear explosion ravaged the centers of the Martian civilization and destroyed the biosphere of the planet. But the story does not end there. This tragedy may explain Fermi's Paradox, the fact that the cosmos, seemingly so fertile and with so many planets suitable for life, is as silent as a graveyard.

278 Pages. 6x9 Paperback. Illustrated. Bibliography. Color Section. $19.95. Code: DOM

BEYOND EINSTEIN'S UNIFIED FIELD
Gravity and Electro-Magnetism Redefined
By John Brandenburg, Ph.D.

Brandenburg reveals the GEM Unification Theory that proves the mathematical and physical interrelation of the forces of gravity and electromagnetism! Brandenburg describes control of space-time geometry through electromagnetism, and states that faster-than-light travel will be possible in the future. Anti-gravity through electromagnetism is possible, which upholds the basic "flying saucer" design utilizing "The Tesla Vortex." Chapters include: Squaring the Circle, Einstein's Final Triumph; A Book of Numbers and Forms; Kepler, Newton and the Sun King; Magnus and Electra; Atoms of Light; Einstein's Glory, Relativity; The Aurora; Tesla's Vortex and the Cliffs of Zeno; The Hidden 5th Dimension; The GEM Unification Theory; Anti-Gravity and Human Flight; The New GEM Cosmos; more. Includes an 8-page color section.

312 Pages. 6x9 Paperback. Illustrated. $18.95. Code: BEUF

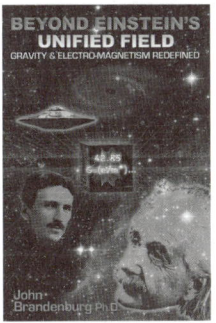

VIMANA:
Flying Machines of the Ancients
by David Hatcher Childress

According to early Sanskrit texts the ancients had several types of airships called vimanas. Like aircraft of today, vimanas were used to fly through the air from city to city; to conduct aerial surveys of uncharted lands; and as delivery vehicles for awesome weapons. David Hatcher Childress, popular *Lost Cities* author and star of the History Channel's long-running show Ancient Aliens, takes us on an astounding investigation into tales of ancient flying machines. In his new book, packed with photos and diagrams, he consults ancient texts and modern stories and presents astonishing evidence that aircraft, similar to the ones we use today, were used thousands of years ago in India, Sumeria, China and other countries. Includes a 24-page color section.

408 Pages. 6x9 Paperback. Illustrated. $22.95. Code: VMA

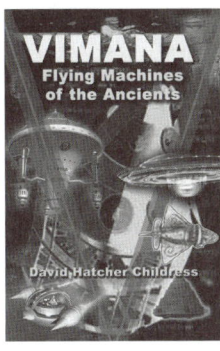

HESS AND THE PENGUINS
The Holocaust, Antarctica and the Strange Case of Rudolf Hess
By Joseph P. Farrell
Farrell looks at Hess' mission to make peace with Britain and get rid of Hitler—even a plot to fly Hitler to Britain for capture! How much did Göring and Hitler know of Rudolf Hess' subversive plot, and what happened to Hess? Why was a doppleganger put in Spandau Prison and then "suicided"? Did the British use an early form of mind control on Hess' double? John Foster Dulles of the OSS and CIA suspected as much. Farrell also uncovers the strange death of Admiral Richard Byrd's son in 1988, about the same time of the death of Hess.
288 Pages. 6x9 Paperback. Illustrated. $19.95. Code: HAPG

HIDDEN FINANCE, ROGUE NETWORKS & SECRET SORCERY
The Fascist International, 9/11, & Penetrated Operations
By Joseph P. Farrell
Farrell investigates the theory that there were not *two* levels to the 9/11 event, but *three*. He says that the twin towers were downed by the force of an exotic energy weapon, one similar to the Tesla energy weapon suggested by Dr. Judy Wood, and ties together the tangled web of missing money, secret technology and involvement of portions of the Saudi royal family. Farrell unravels the many layers behind the 9-11 attack, layers that include the Deutschebank, the Bush family, the German industrialist Carl Duisberg, Saudi Arabian princes and the energy weapons developed by Tesla before WWII.
296 Pages. 6x9 Paperback. Illustrated. $19.95. Code: HFRN

THRICE GREAT HERMETICA & THE JANUS AGE
By Joseph P. Farrell
What do the Fourth Crusade, the exploration of the New World, secret excavations of the Holy Land, and the pontificate of Innocent the Third all have in common? Answer: Venice and the Templars. What do they have in common with Jesus, Gottfried Leibniz, Sir Isaac Newton, Rene Descartes, and the Earl of Oxford? Answer: Egypt and a body of doctrine known as Hermeticism. The hidden role of Venice and Hermeticism reached far and wide, into the plays of Shakespeare (a.k.a. Edward DeVere, Earl of Oxford), into the quest of the three great mathematicians of the Early Enlightenment for a lost form of analysis, and back into the end of the classical era, to little known Egyptian influences at work during the time of Jesus.
354 Pages. 6x9 Paperback. Illustrated. $19.95. Code: TGHJ

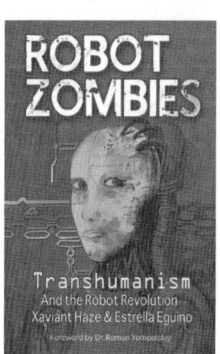

ROBOT ZOMBIES
Transhumanism and the Robot Revolution
By Xaviant Haze and Estrella Eguino,
Technology is growing exponentially and the moment when it merges with the human mind, called "The Singularity," is visible in our imminent future. Science and technology are pushing forward, transforming life as we know it—perhaps even giving humans a shot at immortality. Who will benefit from this? This book examines the history and future of robotics, artificial intelligence, zombies and a Transhumanist utopia/dystopia integrating man with machine. Chapters include: Love, Sex and Compassion—Android Style; Humans Aren't Working Like They Used To; Skynet Rises; Blueprints for Transhumans; Kurzweil's Quest; Nanotech Dreams; Zombies Among Us; Cyborgs (Cylons) in Space; Awakening the Human; more. Color Section.
180 Pages. 6x9 Paperback. Illustrated. $16.95. Code: RBTZ

TRUMPOCALYPSE NOW!
The Triumph of the Conspiracy Spectacle
by Kenn Thomas

Trumpocalypse Now! takes a look at Trump's career as a conspiracy theory celebrity, his trafficking in such notions as birtherism, Islamofascism and 9/11, the conspiracies of the Clinton era, and the JFK assassination. It also examines the controversies of the 2016 election, including the cyber-hacking of the DNC, the Russian involvement and voter fraud. Learn the parapolitcal realities behind the partisan divide and the real ideological underpinnings behind the country's most controversial president. Chapters include: Introduction: Alternative Facts; Conspiracy Celebrity–Trump's TV Career; Birtherism; 9/11 and Islamofascism; Clinton Conspiracies; JFK–Pro-Castro Fakery; Cyber Hacking the DNC; The Russian Connection; Votescam; Conclusion: Alternative Theories; more.

6x9 Paperback. 380 Pages. Illustrated. $16.95. Code: TRPN

MIND CONTROL, OSWALD & JFK
Introduction by Kenn Thomas

In 1969 the strange book *Were We Controlled?* was published which maintained that Lee Harvey Oswald was a special agent who was also a Mind Control subject who had received an implant in 1960. Thomas examines the evidence that Oswald had been an early recipient of the Mind Control implant technology and this startling role in the JFK Assassination. Also: the RHIC-EDOM Mind Control aspects concerning the RFK assassination and the history of implant technology.

256 Pages. 6x9 Paperback. Illustrated. $16.00. Code: MCOJ

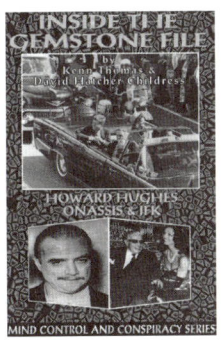

INSIDE THE GEMSTONE FILE
Howard Hughes, Onassis & JFK
By Kenn Thomas & David Childress

Here is the low-down on the most famous underground document ever circulated. Photocopied and distributed for over 20 years, the Gemstone File is the story of Bruce Roberts, the inventor of the synthetic ruby widely used in laser technology today, and his relationship with the Howard Hughes Company and ultimately with Aristotle Onassis, the Mafia, and the CIA. Hughes kidnapped and held a drugged-up prisoner for 10 years; Onassis and his role in the Kennedy Assassination; how the Mafia ran corporate America in the 1960s; more.

320 Pages. 6x9 Paperback. Illustrated. $16.00. Code: IGF

ADVENTURES OF A HASHISH SMUGGLER
by Henri de Monfreid

Nobleman, writer, adventurer and inspiration for the swashbuckling gun runner in the *Adventures of Tintin*, Henri de Monfreid lived by his own account "a rich, restless, magnificent life" as one of the great travelers of his or any age. The son of a French artist who knew Paul Gaugin as a child, de Monfreid sought his fortune by becoming a collector and merchant of the fabled Persian Gulf pearls. He was then drawn into the shadowy world of arms trading, slavery, smuggling and drugs. Infamous as well as famous, his name is inextricably linked to the Red Sea and the raffish ports between Suez and Aden in the early years of the twentieth century. De Monfreid (1879 to 1974) had a long life of many adventures around the Horn of Africa where he dodged pirates as well as the authorities.

284 Pages. 6x9 Paperback. $16.95. Illustrated. Code AHS

SECRETS OF THE UNIFIED FIELD
The Philadelphia Experiment, the Nazi Bell, and the Discarded Theory
by Joseph P. Farrell

American and German wartime scientists determined that, while the Unified Field Theory was incomplete, it could nevertheless be engineered. Chapters include: The Meanings of "Torsion"; The Mistake in Unified Field Theories and Their Discarding by Contemporary Physics; Three Routes to the Doomsday Weapon: Quantum Potential, Torsion, and Vortices; Tesla's Meeting with FDR; Arnold Sommerfeld and Electromagnetic Radar Stealth; Electromagnetic Phase Conjugations, Phase Conjugate Mirrors, and Templates; The Unified Field Theory, the Torsion Tensor, and Igor Witkowski's Idea of the Plasma Focus; tons more.
340 pages. 6x9 Paperback. Illustrated. $18.95. Code: SOUF

THE ANTI-GRAVITY HANDBOOK
edited by David Hatcher Childress

The new expanded compilation of material on Anti-Gravity, Free Energy, Flying Saucer Propulsion, UFOs, Suppressed Technology, NASA Cover-ups and more. Highly illustrated with patents, technical illustrations and photos. This revised and expanded edition has more material, including photos of Area 51, Nevada, the government's secret testing facility. This classic on weird science is back in a new format!
230 PAGES. 7x10 PAPERBACK. ILLUSTRATED. $16.95. CODE: AGH

ANTI–GRAVITY & THE WORLD GRID

Is the earth surrounded by an intricate electromagnetic grid network offering free energy? This compilation of material on ley lines and world power points contains chapters on the geography, mathematics, and light harmonics of the earth grid. Learn the purpose of ley lines and ancient megalithic structures located on the grid. Discover how the grid made the Philadelphia Experiment possible. Explore the Coral Castle and many other mysteries, including acoustic levitation, Tesla Shields and scalar wave weaponry. Browse through the section on anti-gravity patents, and research resources.
274 PAGES. 7x10 PAPERBACK. ILLUSTRATED. $14.95. CODE: AGW

ANTI–GRAVITY & THE UNIFIED FIELD
edited by David Hatcher Childress

Is Einstein's Unified Field Theory the answer to all of our energy problems? Explored in this compilation of material is how gravity, electricity and magnetism manifest from a unified field around us. Why artificial gravity is possible; secrets of UFO propulsion; free energy; Nikola Tesla and anti-gravity airships of the 20s and 30s; flying saucers as superconducting whirls of plasma; anti-mass generators; vortex propulsion; suppressed technology; government cover-ups; gravitational pulse drive; spacecraft & more.
240 PAGES. 7x10 PAPERBACK. ILLUSTRATED. $14.95. CODE: AGU

THE TIME TRAVEL HANDBOOK
A Manual of Practical Teleportation & Time Travel
edited by David Hatcher Childress

The Time Travel Handbook takes the reader beyond the government experiments and deep into the uncharted territory of early time travellers such as Nikola Tesla and Guglielmo Marconi and their alleged time travel experiments, as well as the Wilson Brothers of EMI and their connection to the Philadelphia Experiment—the U.S. Navy's forays into invisibility, time travel, and teleportation. Childress looks into the claims of time travelling individuals, and investigates the unusual claim that the pyramids on Mars were built in the future and sent back in time. A highly visual, large format book, with patents, photos and schematics. Be the first on your block to build your own time travel device!
316 PAGES. 7x10 PAPERBACK. ILLUSTRATED. $16.95. CODE: TTH

ANCIENT ALIENS ON THE MOON
By Mike Bara
What did NASA find in their explorations of the solar system that they may have kept from the general public? How ancient really are these ruins on the Moon? Using official NASA and Russian photos of the Moon, Bara looks at vast cityscapes and domes in the Sinus Medii region as well as glass domes in the Crisium region. Bara also takes a detailed look at the mission of Apollo 17 and the case that this was a salvage mission, primarily concerned with investigating an opening into a massive hexagonal ruin near the landing site. Chapters include: The History of Lunar Anomalies; The Early 20th Century; Sinus Medii; To the Moon Alice!; Mare Crisium; Yes, Virginia, We Really Went to the Moon; Apollo 17; more. Tons of photos of the Moon examined for possible structures and other anomalies.
248 Pages. 6x9 Paperback. Illustrated.. $19.95. Code: AAOM

ANCIENT ALIENS ON MARS
By Mike Bara
Bara brings us this lavishly illustrated volume on alien structures on Mars. Was there once a vast, technologically advanced civilization on Mars, and did it leave evidence of its existence behind for humans to find eons later? Did these advanced extraterrestrial visitors vanish in a solar system wide cataclysm of their own making, only to make their way to Earth and start anew? Was Mars once as lush and green as the Earth, and teeming with life? Chapters include: War of the Worlds; The Mars Tidal Model; The Death of Mars; Cydonia and the Face on Mars; The Monuments of Mars; The Search for Life on Mars; The True Colors of Mars and The Pathfinder Sphinx; more. Color section.
252 Pages. 6x9 Paperback. Illustrated. $19.95. Code: AMAR

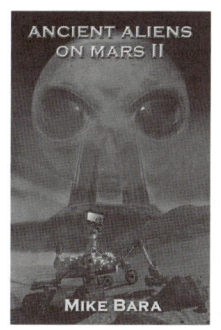

ANCIENT ALIENS ON MARS II
By Mike Bara
Using data acquired from sophisticated new scientific instruments like the Mars Odyssey THEMIS infrared imager, Bara shows that the region of Cydonia overlays a vast underground city full of enormous structures and devices that may still be operating. He peels back the layers of mystery to show images of tunnel systems, temples and ruins, and exposes the sophisticated NASA conspiracy designed to hide them. Bara also tackles the enigma of Mars' hollowed out moon Phobos, and exposes evidence that it is artificial. Long-held myths about Mars, including claims that it is protected by a sophisticated UFO defense system, are examined. Data from the Mars rovers Spirit, Opportunity and Curiosity are examined; everything from fossilized plants to mechanical debris is exposed in images taken directly from NASA's own archives.
294 Pages. 6x9 Paperback. Illustrated. $19.95. Code: AAM2

ANCIENT TECHNOLOGY IN PERU & BOLIVIA
By David Hatcher Childress
Childress speculates on the existence of a sunken city in Lake Titicaca and reveals new evidence that the Sumerians may have arrived in South America 4,000 years ago. He demonstrates that the use of "keystone cuts" with metal clamps poured into them to secure megalithic construction was an advanced technology used all over the world, from the Andes to Egypt, Greece and Southeast Asia. He maintains that only power tools could have made the intricate articulation and drill holes found in extremely hard granite and basalt blocks in Bolivia and Peru, and that the megalith builders had to have had advanced methods for moving and stacking gigantic blocks of stone, some weighing over 100 tons.
340 Pages. 6x9 Paperback. Illustrated.. $19.95 Code: ATP

THE COSMIC WAR
Interplanetary Warfare, Modern Physics, and Ancient Texts
By Joseph P. Farrell
There is ample evidence across our solar system of catastrophic events. The asteroid belt may be the remains of an exploded planet! The known planets are scarred from incredible impacts, and teeter in their orbits due to causes heretofore inadequately explained. Included: The history of the Exploded Planet hypothesis, and what mechanism can actually explode a planet. The role of plasma cosmology, plasma physics and scalar physics. The ancient texts telling of such destructions: from Sumeria (Tiamat's destruction by Marduk), Egypt (Edfu and the Mars connections), Greece (Saturn's role in the War of the Titans) and the ancient Americas.
436 Pages. 6x9 Paperback. Illustrated.. $18.95. Code: COSW

THE GRID OF THE GODS
The Aftermath of the Cosmic War & the Physics of the Pyramid Peoples
By Joseph P. Farrell with Scott D. de Hart
Farrell looks at Ashlars and Engineering; Anomalies at the Temples of Angkor; The Ancient Prime Meridian: Giza; Transmitters, Nazis and Geomancy; the Lithium-7 Mystery; Nazi Transmitters and the Earth Grid; The Master Plan of a Hidden Elite; Moving and Immoveable Stones; Uncountable Stones and Stones of the Giants and Gods; Gateway Traditions; The Grid and the Ancient Elite; Finding the Center of the Land; The Ancient Catastrophe, the Very High Civilization, and the Post-Catastrophe Elite; Tiahuanaco and the Puma Punkhu Paradox: Ancient Machining; The Black Brotherhood and Blood Sacrifices; The Gears of Giza: the Center of the Machine; Alchemical Cosmology and Quantum Mechanics in Stone; tons more.
436 Pages. 6x9 Paperback. Illustrated. $19.95. Code: GOG

THE SS BROTHERHOOD OF THE BELL
The Nazis' Incredible Secret Technology
by Joseph P. Farrell
In 1945, a mysterious Nazi secret weapons project code-named "The Bell" left its underground bunker in lower Silesia, along with all its project documentation, and a four-star SS general named Hans Kammler. Taken aboard a massive six engine Junkers 390 ultra-long range aircraft, "The Bell," Kammler, and all project records disappeared completely, along with the gigantic aircraft. It is thought to have flown to America or Argentina. What was "The Bell"? What new physics might the Nazis have discovered with it? How far did the Nazis go after the war to protect the advanced energy technology that it represented?
456 pages. 6x9 Paperback. Illustrated. $16.95. Code: SSBB

MAPS OF THE ANCIENT SEA KINGS
Evidence of Advanced Civilization in the Ice Age
by Charles H. Hapgood
Charles Hapgood has found the evidence in the Piri Reis Map that shows Antarctica, the Hadji Ahmed map, the Oronteus Finaeus and other amazing maps. Hapgood concluded that these maps were made from more ancient maps from the various ancient archives around the world, now lost. Not only were these unknown people more advanced in mapmaking than any people prior to the 18th century, it appears they mapped all the continents. The Americas were mapped thousands of years before Columbus. Antarctica was mapped when its coasts were free of ice!
316 PAGES. 7x10 PAPERBACK. ILLUSTRATED. $19.95. CODE: MASK

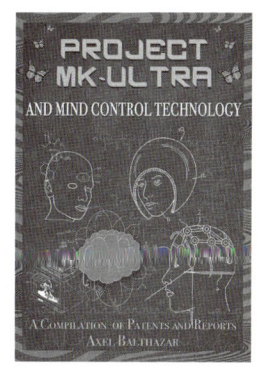

PROJECT MK-ULTRA AND MIND CONTROL TECHNOLOGY
A Compilation of Patents and Reports
By Axel Balthazar
This book is a compilation of the government's documentation on MK-Ultra, the CIA's mind control experimentation on unwitting human subjects, as well as over 150 patents pertaining to artificial telepathy (voice-to-skull technology), behavior modification through radio frequencies, directed energy weapons, electronic monitoring, implantable nanotechnology, brain wave manipulation, nervous system manipulation, neuroweapons, psychological warfare, satellite terrorism, subliminal messaging, and more. A must-have reference guide for targeted individuals and anyone interested in the subject of mind control technology.
384 pages. 7x10 Paperback. Illustrated. $19.95. Code: PMK

ANCIENT ALIENS & SECRET SOCIETIES
By Mike Bara
Did ancient "visitors"—of extraterrestrial origin—come to Earth long, long ago and fashion man in their own image? Were the science and secrets that they taught the ancients intended to be a guide for all humanity to the present era? Bara establishes the reality of the catastrophe that jolted the human race, and traces the history of secret societies from the priesthood of Amun in Egypt to the Templars in Jerusalem and the Scottish Rite Freemasons. Bara also reveals the true origins of NASA and exposes the bizarre triad of secret societies in control of that agency since its inception. Chapters include: Out of the Ashes; From the Sky Down; Ancient Aliens?; The Dawn of the Secret Societies; The Fractures of Time; Into the 20th Century; The Wink of an Eye; more.
288 Pages. 6x9 Paperback. Illustrated. $19.95. Code: AASS

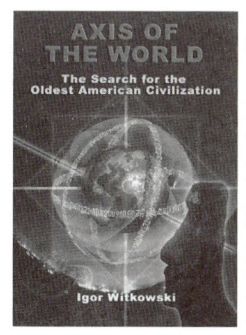

AXIS OF THE WORLD
The Search for the Oldest American Civilization
by Igor Witkowski
Polish author Witkowski's research reveals remnants of a high civilization that was able to exert its influence on almost the entire planet, and did so with full consciousness. Sites around South America show that this was not just one of the places influenced by this culture, but a place where they built their crowning achievements. Easter Island, in the southeastern Pacific, constitutes one of them. The Rongo-Rongo language that developed there points westward to the Indus Valley. Taken together, the facts presented by Witkowski provide a fresh, new proof that an antediluvian, great civilization flourished several millennia ago.
220 pages. 6x9 Paperback. Illustrated. References. $18.95. Code: AXOW

LEY LINE & EARTH ENERGIES
An Extraordinary Journey into the Earth's Natural Energy System
by David Cowan & Chris Arnold
The mysterious standing stones, burial grounds and stone circles that lace Europe, the British Isles and other areas have intrigued scientists, writers, artists and travellers through the centuries. How do ley lines work? How did our ancestors use Earth energy to map their sacred sites and burial grounds? How do ghosts and poltergeists interact with Earth energy? How can Earth spirals and black spots affect our health? This exploration shows how natural forces affect our behavior, how they can be used to enhance our health and well being.
368 PAGES. 6x9 PAPERBACK. ILLUSTRATED. $18.95. CODE: LLEE

SAUCERS, SWASTIKAS AND PSYOPS
A History of a Breakaway Civilization
By Joseph P. Farrell
Farrell discusses SS Commando Otto Skorzeny; George Adamski; the alleged Hannebu and Vril craft of the Third Reich; The Strange Case of Dr. Hermann Oberth; Nazis in the US and their connections to "UFO contactees"; The Memes—an idea or behavior spread from person to person within a culture—are Implants. Chapters include: The Nov. 20, 1952 Contact: The Memes are Implants; The Interplanetary Federation of Brotherhood; Adamski's Technological Descriptions and Another ET Message: The Danger of Weaponized Gravity; Adamski's Retro-Looking Saucers, and the Nazi Saucer Myth; Dr. Oberth's 1968 Statements on UFOs and Extraterrestrials; more.
272 Pages. 6x9 Paperback. Illustrated. $19.95. Code: SSPY

LBJ AND THE CONSPIRACY TO KILL KENNEDY
By Joseph P. Farrell
Farrell says that a coalescence of interests in the military industrial complex, the CIA, and Lyndon Baines Johnson's powerful and corrupt political machine in Texas led to the events culminating in the assassination of JFK. Chapters include: Oswald, the FBI, and the CIA: Hoover's Concern of a Second Oswald; Oswald and the Anti-Castro Cubans; The Mafia; Hoover, Johnson, and the Mob; The FBI, the Secret Service, Hoover, and Johnson; The CIA and "Murder Incorporated"; Ruby's Bizarre Behavior; The French Connection and Permindex; Big Oil; The Dead Witnesses: Guy Bannister, Jr., Mary Pinchot Meyer, Rose Cheramie, Dorothy Killgallen, Congressman Hale Boggs; LBJ and the Planning of the Texas Trip; LBJ: A Study in Character, Connections, and Cabals; LBJ and the Aftermath: Accessory After the Fact; The Requirements of Coups D'État; more.
342 Pages. 6x9 Paperback. $19.95 Code: LCKK

THE TESLA PAPERS
Nikola Tesla on Free Energy &
Wireless Transmission of Power
by Nikola Tesla, edited by David Hatcher Childress
David Hatcher Childress takes us into the incredible world of Nikola Tesla and his amazing inventions. Tesla's fantastic vision of the future, including wireless power, anti-gravity, free energy and highly advanced solar power. Also included are some of the papers, patents and material collected on Tesla at the Colorado Springs Tesla Symposiums, including papers on: •The Secret History of Wireless Transmission •Tesla and the Magnifying Transmitter •Design and Construction of a Half-Wave Tesla Coil •Electrostatics: A Key to Free Energy •Progress in Zero-Point Energy Research •Electromagnetic Energy from Antennas to Atoms
325 PAGES. 8x10 PAPERBACK. ILLUSTRATED. $16.95. CODE: TTP

COVERT WARS & THE CLASH OF CIVILIZATIONS
UFOs, Oligarchs and Space Secrecy
By Joseph P. Farrell
Farrell's customary meticulous research and sharp analysis blow the lid off of a worldwide web of nefarious financial and technological control that very few people even suspect exists. He elaborates on the advanced technology that they took with them at the "end" of World War II and shows how the breakaway civilizations have created a huge system of hidden finance with the involvement of various banks and financial institutions around the world. He investigates the current space secrecy that involves UFOs, suppressed technologies and the hidden oligarchs who control planet earth for their own gain and profit.
358 Pages. 6x9 Paperback. Illustrated. $19.95. Code: CWCC

HITLER'S SUPPRESSED AND STILL-SECRET WEAPONS, SCIENCE AND TECHNOLOGY
by Henry Stevens
In the closing months of WWII the Allies assembled mind-blowing intelligence reports of supermetals, electric guns, and ray weapons able to stop the engines of Allied aircraft—in addition to feared x-ray and laser weaponry. Chapters include: The Kammler Group; German Flying Disc Update; The Electromagnetic Vampire; Liquid Air; Synthetic Blood; German Free Energy Research; German Atomic Tests; The Fuel-Air Bomb; Supermetals; Red Mercury; Means to Stop Engines, more.
335 Pages. 6x9 Paperback. Illustrated. $19.95. Code: HSSW

PRODIGAL GENIUS
The Life of Nikola Tesla
by John J. O'Neill
This special edition of O'Neill's book has many rare photographs of Tesla and his most advanced inventions. Tesla's eccentric personality gives his life story a strange romantic quality. He made his first million before he was forty, yet gave up his royalties in a gesture of friendship, and died almost in poverty. Tesla could see an invention in 3-D, from every angle, within his mind, before it was built; how he refused to accept the Nobel Prize; his friendships with Mark Twain, George Westinghouse and competition with Thomas Edison. Tesla is revealed as a figure of genius whose influence on the world reaches into the far future. Deluxe, illustrated edition.
408 pages. 6x9 Paperback. Illustrated. Bibliography. $18.95. Code: PRG

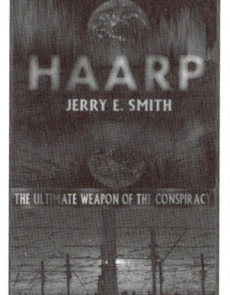

HAARP
The Ultimate Weapon of the Conspiracy
by Jerry Smith
The HAARP project in Alaska is one of the most controversial projects ever undertaken by the U.S. Government. At at worst, HAARP could be the most dangerous device ever created, a futuristic technology that is everything from super-beam weapon to world-wide mind control device. Topics include Over-the-Horizon Radar and HAARP, Mind Control, ELF and HAARP, The Telsa Connection, The Russian Woodpecker, GWEN & HAARP, Earth Penetrating Tomography, Weather Modification, Secret Science of the Conspiracy, more. Includes the complete 1987 Eastlund patent for his pulsed super-weapon that he claims was stolen by the HAARP Project.
256 pages. 6x9 Paperback. Illustrated. Bib. $14.95. Code: HARP

WEATHER WARFARE
The Military's Plan to Draft Mother Nature
by Jerry E. Smith
Weather modification in the form of cloud seeding to increase snow packs in the Sierras or suppress hail over Kansas is now an everyday affair. Underground nuclear tests in Nevada have set off earthquakes. A Russian company has been offering to sell typhoons (hurricanes) on demand since the 1990s. Scientists have been searching for ways to move hurricanes for over fifty years. In the same amount of time we went from the Wright Brothers to Neil Armstrong. Hundreds of environmental and weather modifying technologies have been patented in the United States alone – and hundreds more are being developed in civilian, academic, military and quasi-military laboratories around the world *at this moment!* Numerous ongoing military programs do inject aerosols at high altitude for communications and surveillance operations.
304 Pages. 6x9 Paperback. Illustrated. Bib. $18.95. Code: WWAR

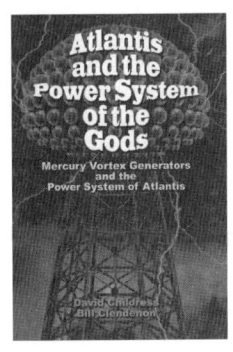

ATLANTIS & THE POWER SYSTEM OF THE GODS
by David Hatcher Childress and Bill Clendenon
Childress' fascinating analysis of Nikola Tesla's broadcast system in light of Edgar Cayce's "Terrible Crystal" and the obelisks of ancient Egypt and Ethiopia. Includes: Atlantis and its crystal power towers that broadcast energy; how these incredible power stations may still exist today; inventor Nikola Tesla's nearly identical system of power transmission; Mercury Proton Gyros and mercury vortex propulsion; more. Richly illustrated, and packed with evidence that Atlantis not only existed—it had a worldwide energy system more sophisticated than ours today.
246 PAGES. 6x9 PAPERBACK. ILLUSTRATED. $15.95. CODE: APSG

TECHNOLOGY OF THE GODS
The Incredible Sciences of the Ancients
by David Hatcher Childress
Childress looks at the technology that was allegedly used in Atlantis and the theory that the Great Pyramid of Egypt was originally a gigantic power station. He examines tales of ancient flight and the technology that it involved; how the ancients used electricity; megalithic building techniques; the use of crystal lenses and the fire from the gods; evidence of various high tech weapons in the past, including atomic weapons; ancient metallurgy and heavy machinery; the role of modern inventors such as Nikola Tesla in bringing ancient technology back into modern use; impossible artifacts; and more.
356 PAGES. 6x9 PAPERBACK. ILLUSTRATED. BIBLIOGRAPHY. $16.95. CODE: TGOD

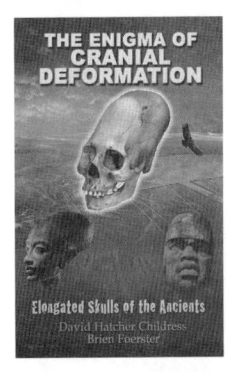

THE ENIGMA OF CRANIAL DEFORMATION
Elongated Skulls of the Ancients
By David Hatcher Childress and Brien Foerster
In a book filled with over a hundred astonishing photos and a color photo section, Childress and Foerster take us to Peru, Bolivia, Egypt, Malta, China, Mexico and other places in search of strange elongated skulls and other cranial deformation. The puzzle of why diverse ancient people—even on remote Pacific Islands—would use head-binding to create elongated heads is mystifying. Where did they even get this idea? Did some people naturally look this way—with long narrow heads? Were they some alien race? Were they an elite race that roamed the entire planet? Why do anthropologists rarely talk about cranial deformation and know so little about it? Color Section.
250 Pages. 6x9 Paperback. Illustrated. $19.95. Code: ECD

ARK OF GOD
The Incredible Power of the Ark of the Covenant
By David Hatcher Childress
Childress takes us on an incredible journey in search of the truth about (and science behind) the fantastic biblical artifact known as the Ark of the Covenant. This object made by Moses at Mount Sinai—part wooden-metal box and part golden statue—had the power to create "lightning" to kill people, and also to fly and lead people through the wilderness. The Ark of the Covenant suddenly disappears from the Bible record and what happened to it is not mentioned. Was it hidden in the underground passages of King Solomon's temple and later discovered by the Knights Templar? Was it taken through Egypt to Ethiopia as many Coptic Christians believe? Childress looks into hidden history, astonishing ancient technology, and a 3,000-year-old mystery that continues to fascinate millions of people today. Color section.
420 Pages. 6x9 Paperback. Illustrated. $22.00 Code: AOG

LIQUID CONSPIRACY 2:
The CIA, MI6 & Big Pharma's War on Psychedelics
By Xaviant Haze
Underground author Xaviant Haze looks into the CIA and its use of LSD as a mind control drug; at one point every CIA officer had to take the drug and endure mind control tests and interrogations to see if the drug worked as a "truth serum." Chapters include: The Pioneers of Psychedelia; The United Kingdom Mellows Out: The MI5, MDMA and LSD; Taking it to the Streets. LSD becomes Acid, Great Works of Art Inspired and Influenced by Acid; Scapolamine: The CIA's Ultimate Truth Serum; Mind Control, the Death of Music and the Meltdown of the Masses; Big Pharma's War on Psychedelics; The Healing Powers of Psychedelic Medicine; tons more.
240 pages. 6x9 Paperback. Illustrated. $19.95. Code: LQC2

TAPPING THE ZERO POINT ENERGY
Free Energy & Anti-Gravity in Today's Physics
by Moray B. King
King explains how free energy and anti-gravity are possible. The theories of the zero point energy maintain there are tremendous fluctuations of electrical field energy imbedded within the fabric of space. This book tells how, in the 1930s, inventor T. Henry Moray could produce a fifty kilowatt "free energy" machine; how an electrified plasma vortex creates anti-gravity; how the Pons/Fleischmann "cold fusion" experiment could produce tremendous heat without fusion; and how certain experiments might produce a gravitational anomaly.
180 PAGES. 5x8 PAPERBACK. ILLUSTRATED. $12.95. CODE: TAP

QUEST FOR ZERO-POINT ENERGY
Engineering Principles for "Free Energy"
by Moray B. King
King expands, with diagrams, on how free energy and anti-gravity are possible. The theories of zero point energy maintain there are tremendous fluctuations of electrical field energy embedded within the fabric of space. King explains the following topics: TFundamentals of a Zero-Point Energy Technology; Vacuum Energy Vortices; The Super Tube; Charge Clusters: The Basis of Zero-Point Energy Inventions; Vortex Filaments, Torsion Fields and the Zero-Point Energy; Transforming the Planet with a Zero-Point Energy Experiment; Dual Vortex Forms: The Key to a Large Zero-Point Energy Coherence. Packed with diagrams, patents and photos.
224 PAGES. 6x9 PAPERBACK. ILLUSTRATED. $14.95. CODE: QZPE

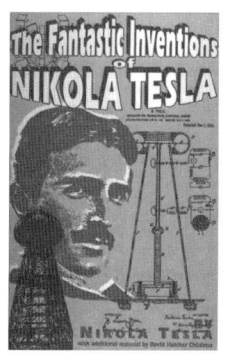

THE FANTASTIC INVENTIONS OF NIKOLA TESLA
by Nikola Tesla with David Hatcher Childress
This book is a readable compendium of patents, diagrams, photos and explanations of the many incredible inventions of the originator of the modern era of electrification. In Tesla's own words are such topics as wireless transmission of power, death rays, and radio-controlled airships. In addition, rare material on a secret city built at a remote jungle site in South America by one of Tesla's students, Guglielmo Marconi. Marconi's secret group claims to have built flying saucers in the 1940s and to have gone to Mars in the early 1950s! Incredible photos of these Tesla craft are included. •His plan to transmit free electricity into the atmosphere. •How electrical devices would work using only small antennas. •Why unlimited power could be utilized anywhere on earth. •How radio and radar technology can be used as death-ray weapons in Star Wars.
342 PAGES. 6x9 PAPERBACK. ILLUSTRATED. $16.95. CODE: FINT

ORDER FORM

10% Discount When You Order 3 or More Items!

One Adventure Place
P.O. Box 74
Kempton, Illinois 60946
United States of America
Tel.: 815-253-6390 • Fax: 815-253-6300
Email: auphq@frontiernet.net
http://www.adventuresunlimitedpress.com

ORDERING INSTRUCTIONS

- ✓ Remit by USD$ Check, Money Order or Credit Card
- ✓ Visa, Master Card, Discover & AmEx Accepted
- ✓ Paypal Payments Can Be Made To: info@wexclub.com
- ✓ Prices May Change Without Notice
- ✓ 10% Discount for 3 or More Items

SHIPPING CHARGES

United States
- ✓ Postal Book Rate { $4.50 First Item / 50¢ Each Additional Item
- ✓ POSTAL BOOK RATE Cannot Be Tracked! Not responsible for non-delivery.
- ✓ Priority Mail { $6.00 First Item / $2.00 Each Additional Item
- ✓ UPS { $7.00 First Item / $1.50 Each Additional Item

NOTE: UPS Delivery Available to Mainland USA Only

Canada
- ✓ Postal Air Mail { $15.00 First Item / $2.50 Each Additional Item
- ✓ Personal Checks or Bank Drafts MUST BE US$ and Drawn on a US Bank
- ✓ Canadian Postal Money Orders OK
- ✓ Payment MUST BE US$

All Other Countries
- ✓ Sorry, No Surface Delivery!
- ✓ Postal Air Mail { $19.00 First Item / $6.00 Each Additional Item
- ✓ Checks and Money Orders MUST BE US$ and Drawn on a US Bank or branch.
- ✓ Paypal Payments Can Be Made in US$ To: info@wexclub.com

SPECIAL NOTES

- ✓ RETAILERS: Standard Discounts Available
- ✓ BACKORDERS: We Backorder all Out-of-Stock Items Unless Otherwise Requested
- ✓ PRO FORMA INVOICES: Available on Request
- ✓ DVD Return Policy: Replace defective DVDs only

ORDER ONLINE AT: www.adventuresunlimitedpress.com

10% Discount When You Order 3 or More Items!

Please check: ✓

☐ This is my first order   ☐ I have ordered before

| Name |
| Address |
| City |
| State/Province | Postal Code |
| Country |
| Phone: Day | Evening |
| Fax | Email |

Item Code	Item Description	Qty	Total

Subtotal ▶
Please check: ✓   Less Discount-10% for 3 or more items ▶

☐ Postal-Surface   Balance ▶
☐ Postal-Air Mail Illinois Residents 6.25% Sales Tax ▶
   (Priority in USA)   Previous Credit ▶
☐ UPS   Shipping ▶
   (Mainland USA only) Total (check/MO in USD$ only) ▶

☐ Visa/MasterCard/Discover/American Express

Card Number:

Expiration Date: Security Code:

✓ SEND A CATALOG TO A FRIEND: